Piece-wise and Max-Type
Difference Equations

Piece-wise and Max-Type Difference Equations

Periodic and Eventually Periodic Solutions

Michael A. Radin

CRC Press

Taylor & Francis Group

Boca Raton London New York

CRC Press is an imprint of the
Taylor & Francis Group, an **informa** business

A CHAPMAN & HALL BOOK

First edition published 2021
by CRC Press
6000 Broken Sound Parkway NW, Suite 300, Boca Raton, FL 33487-2742

and by CRC Press
2 Park Square, Milton Park, Abingdon, Oxon, OX14 4RN

First issued in paperback 2022

Library of Congress Cataloging-in-Publication Data
Names: Radin, Michael A. (Michael Alexander), author.
Title: Piece-wise and Max-type difference equations : periodic and
eventually periodic solutions / Michael A. Radin.
Description: First edition. | Boca Raton, FL : CRC Press, 2020. |
Includes bibliographical references and index.
Identifiers: LCCN 2020017699 (print) | LCCN 2020017700 (ebook) |
ISBN 9781138313507 (hardback) | ISBN 9780429457616 (ebook)
Subjects: LCSH: Difference equations.
Classification: LCC QA431 .R3256 2020 (print) | LCC QA431 (ebook) |
DDC 515/.625—dc23
LC record available at https://lccn.loc.gov/2020017699
LC ebook record available at https://lccn.loc.gov/2020017700

ISBN: 978-1-138-31350-7 (hbk)
ISBN: 978-0-367-55102-5 (pbk)
ISBN: 978-0-429-45761-6 (ebk)

DOI: 10.1201/9780429457616

Typeset in CMR10
by codeMantra

Visit the Taylor & Francis Web site at
http://www.taylorandfrancis.com

and the CRC Press Web site at
http//www.crcpress.com

Contents

Preface

Periodic traits and patterns frequently emerge in various designs and in natural phenomena. Analyzing periodic characteristics is essential for the development and enhancement of our intuition, analytical, inductive and deductive reasoning skills. Periodic features often arise while we are analyzing weather patterns and natural phenomena, while studying music, while studying a foreign language, while learning computer programming, while observing traffic patterns, while designing engineering structures and while detecting and processing signals.

The aim of this interdisciplinary book is to detect and describe formations of periodic and eventually periodic patterns analytically and graphically by remitting the corresponding questions:

- An even-ordered periodic cycle or an odd-ordered periodic cycle?

- Uniqueness of periodic cycles?

- The relationship between the neighboring terms of the cycle?

- Positive, negative or alternating periodic cycles?

- Shapes of periodic cycles such as ascending, descending, triangular shaped, step shaped, trapezoidal shaped, etc?

- What are the necessary and sufficient conditions for the existence of periodic cycles and transient terms?

- When do eventually periodic solutions with transient terms emerge and why?

- How do we describe the transient terms' patterns? In one sequence or in two or more sub-sequences?

I invite you to the discovery journey in deciphering discrete periodic patterns of first and higher order difference equations analytically and graphically. We will study the periodic traits of First-Order Linear Difference Equations, Riccati Difference Equations, First-Order Piece-wise Difference Equations and Max-type Difference Equations and examine various applications. Our plan is to develop an inductive intuition that will guide us to detecting specific formulations of periodic cycles and eventually periodic cycles, expanding our knowledge and comprehension of assorted patterns and devising theorems after numerous repetitive-type examples that will be provided in each chapter.

Michael A. Radin

Acknowledgments

First of all, I would like to take the opportunity to thank the CRC Press staff for their support, encouragement, and beneficial guidance while devising new ideas and for keeping me on the right time track. Their encouragements certainly lead me in the new innovative directions with new practices and formulations of concepts. Their suggestions were very valuable with the textbook's structure such as introduction of new definitions, graphical representations of concepts, additional examples of new concepts and applying the definitions and principles starting from the introduction chapter throughout the textbook.

Second, I would also like to thank my colleague Olga A. Orlova from Munich Technical University for her artistic help with numerous diagrams. Olga emphasized several mistakes that she detected after meticulously checking the examples in each section and in the end-of-chapter exercises. In addition, Olga suggested to include specific additional examples of configuration of assorted periodic patterns and supplemental examples rendering transient terms and piece-wise sequences. She encouraged to emphasize the use of piece-wise sequences in the formulations of specific patterns throughout the textbook. In closing, Olga said, "This will be one of many interdisciplinary textbooks ahead".

Furthermore, I would like to thank my colleague Candace Kent from Virginia Commonwealth University for her guidance while discovering the boundedness and periodic character of Non-Autonomous Max-Type Difference Equations. She pointed me in the new direction of study on Max-Type Difference Equations. I would also like to thank my colleague Inese Bula from the University of Latvia for inviting me to work on a Piece-Wise Difference Equation that renders very diverse periodic traits with sensitivity on initial condition. Her guidance lead me to new discoveries and innovations.

Finally, I would like to thank my parents Alexander and Shulamit for encouraging me to write textbooks, for their support with the textbook's content and for persuading me to continue writing future textbooks.

Author

Michael A. Radin earned his Ph.D. at the University of Rhode Island in 2001 and is currently an associate professor of mathematics at the Rochester Institute of Technology. Dr. Radin started his journey analyzing difference equations with periodic and eventually periodic solutions as part of his Ph.D. dissertation and has numerous publications on boundedness and periodic nature of solutions of rational difference equations, max-type difference equations and piece-wise difference equations. Dr. Radin published several papers together with his post- and undergraduate students at RIT and with students and colleagues from Riga Technical University, the University of Latvia and Yaroslavl State University.

Dr. Radin also has publications in neural networking, modelling extinct civilizations, modelling human emotions and papers on international pedagogical innovations. In addition, Dr. Radin organized numerous sessions on difference equations and applications at the annual **American Mathematical Society** meetings and presents his research at international conferences such as Conference on Mathematical Modelling and Analysis and Volga Neuroscience Meeting. Recently Dr. Radin published two manuscripts on international pedagogy and has been invited as a keynote speaker at several international and interdisciplinary conferences. Dr. Radin taught courses and conducted seminars on these related topics during his spring 2009 sabbatical at the Aegean University in Greece and during his spring 2016 sabbatical at Riga Technical University in Latvia. Dr. Radin's aim is to inspire students to learn.

Recently, Dr. Radin had the opportunity to implement the hands-on teaching and learning style in the courses that he regularly teaches at RIT and during his sabbatical in Latvia during the spring 2016 semester. This method confirmed to work very successfully for him and his students, kept the students stimulated and improved their course performance [11,12]. Therefore, the hands-on teaching and learning style is the intent of this book by providing numerous repetitive-type examples. In fact, several repetitive-type examples will develop our intuition on patterns' recognition and help see the wider spectrum on how concepts relate to each other and will lead to formulation of theorems and their proofs. This will be an essential technique to understand the proof by an induction method that will be used to describe numerous results throughout the book.

During his spare time, Dr. Radin spends time outdoors and is an avid land-scape photographer. In addition, Dr. Radin is an active poet and has several published poems in the **LeMot Juste**. Furthermore, Dr. Radin published an article titled **"Re-photographing the Baltic Sea Scenery in Liepaja: Why photograph the same scenery multiple times"** in the *Journal of Humanities and Arts 2018*. Dr. Radin also recently published the book **"Poetic Landscape Photography"** with *JustFiction Edition 2019*. Spending time outdoors and active landscape photography widens and expands Dr. Radin's understandings of nature's patterns and cadences.

Chapter 1

Introduction

It is essential to study and analyze patterns as we come across them on a daily basis. For instance, we encounter traffic patterns when we are driving, musical patterns when we are learning to play a musical instrument, nature's patterns and weather patterns when we spend time outdoors, foreign languages' patterns, signals' patterns in signal processing, patterns in computer programming and mathematical patterns. We can discover patterns that repeat at the same scale, patterns that repeat at different scales, and alternating patterns. The first example that I would like to share is the repetition of clouds' patterns and grassy dunes at the same scale (Figure 1.1) that we can observe along the Baltic scenery in Latvia:

FIGURE 1.1: Repeated Patterns of Dunes and Clouds at the Same Scale.

The second example is an aerial alpine photograph of the Rocky Mountains (Figure 1.2) rendering patterns repeated at different scales:

FIGURE 1.2: Alpine Patterns Repeated at Different Scales.

The next photograph (Figure 1.3) depicts an alternating patterns of waves in the Gulf of Giga in Latvia:

FIGURE 1.3: Alternating Patterns of Waves in the Gulf of Riga.

The textbook's aims are to get acquainted with difference equations that resemble periodicity, which will then guide us to the study of **Piece-wise Difference Equations** and **Max-Type Difference Equations**. Our goals are to study the periodic character of Piece-wise Difference Equations and Max-Type Difference Equations and compare their similarities and differences. In addition, the intents of this textbook are to familiarize ourselves with difference equations (recursive sequences), to solve them explicitly using inductive reasoning and to determine the patterns and traits of periodic cycles and to inductively understand, assemble and develop patterns after several repetitive-type examples, which will then lead to the discovery of theorems. Further, the textbook also aims to detect alternating patterns, the relationship between the indices from neighbor to neighbor and to compare the similarities and differences between even-ordered periodic cycles and odd-ordered periodic cycles and additional such characteristics. Our primary focus will be on periodic and eventually periodic solutions of **Piece-wise Difference Equations** and **Max-Type Difference Equations**. Throughout the book, we will remit why only specific difference equations exhibit transient behavior. We will abbreviate difference equation as $\Delta.E$.

1.1   Recursive Sequences

We will commence with the introduction of difference equations as recursive sequences. Our aims are to determine a specific pattern of a sequence and express it recursively. We will treat it as an **Initial Value Problem**, where the initial value is the starting value of a sequence. We will emerge with examples by transitioning from neighbor to neighbor by adding. The very first example lists all the positive multiples of 3 starting with 3.

Example 1.1 *Write a recursive formula for:*

$$3, 6, 9, 12, 15, 18, 21, \ldots.$$

Solution: *Observe that we start at 3 and transition from neighbor to neighbor by adding a 3. Thus, recursively and inductively we procure:*

$$x_0 = 3,$$
$$x_0 + 3 = 3 + 3 = 6 = x_1,$$
$$x_1 + 3 = 6 + 3 = 9 = x_2,$$
$$x_2 + 3 = 9 + 3 = 12 = x_3,$$
$$x_3 + 3 = 12 + 3 = 15 = x_4,$$
$$x_4 + 3 = 15 + 3 = 18 = x_5,$$
$$\vdots$$

Thus for all $n \geq 0$:

$$\begin{cases} x_{n+1} = x_n + 3, \\ \quad x_0 = 3. \end{cases}$$

The consequent example will add consecutive positive integers starting at 1.

Example 1.2 *Write a recursive formula for:*

$$1, 3, 6, 10, 15, 21, 28, \ldots.$$

Solution: *Note that we start at 1 and transition to the next term by adding a 2, then add a 3 and so on. Hence, recursively and inductively we acquire:*

$$x_0 = 1,$$
$$x_0 + 2 = 1 + 2 = 3 = x_1,$$
$$x_1 + 3 = 3 + 2 = 6 = x_2,$$
$$x_2 + 4 = 6 + 4 = 10 = x_3,$$
$$x_3 + 5 = 10 + 5 = 15 = x_4,$$
$$x_4 + 6 = 15 + 6 = 21 = x_5,$$
$$\vdots$$

Thus for all $n \geq 0$:

$$\begin{cases} x_{n+1} = x_n + (n + 2), \\ \quad x_0 = 1. \end{cases}$$

The succeeding examples evokes a **Geometric Sequence**.

Example 1.3 *Write a recursive formula for:*

$$4, 8, 16, 32, 64, 128, 256, \ldots.$$

Solution: *Notice that we transition from neighbor to neighbor by multiplying by 2. Therefore, recursively and inductively we get:*

$$x_0 = 4,$$
$$x_0 \cdot 2 = 4 \cdot 2 = 8 = x_1,$$
$$x_1 \cdot 2 = 8 \cdot 2 = 16 = x_2,$$
$$x_2 \cdot 2 = 16 \cdot 2 = 32 = x_3,$$
$$x_3 \cdot 2 = 32 \cdot 2 = 64 = x_4,$$
$$x_4 \cdot 2 = 64 \cdot 2 = 128 = x_5,$$
$$\vdots$$

Thus for all $n \geq 0$:

$$\begin{cases} x_{n+1} = 2x_n, \\ \quad x_0 = 4. \end{cases}$$

The upcoming example renders the **Factorial Pattern**.

Example 1.4 *Write a recursive formula for:*

$$1, 2, 6, 24, 120, 720, 5040, \ldots.$$

Solution: *We obtain the following iterative pattern:*

$$x_0 = 1,$$
$$x_0 \cdot 2 = 1 \cdot 2 = 2 = x_1,$$
$$x_1 \cdot 3 = 2 \cdot 3 = 6 = x_2,$$
$$x_2 \cdot 4 = 6 \cdot 4 = 24 = x_3,$$
$$x_3 \cdot 5 = 24 \cdot 5 = 120 = x_4,$$
$$x_4 \cdot 6 = 120 \cdot 6 = 720 = x_5,$$
$$\vdots$$

Thus for all $n \geq 0$:

$$\begin{cases} x_{n+1} = (n+2) \cdot x_n, \\ \quad x_0 = 1. \end{cases}$$

1.2 Order and Explicit Solution of a $\Delta.E.$

We will emerge with difference equations of different orders:

(i) $x_{n+1} = \frac{1}{x_n}, \quad n = 0, 1, \ldots.$

(ii) $x_{n+2} = \max\left\{\frac{1}{x_{n+1}}, \frac{2}{x_n}\right\}, \quad n = 0, 1, \ldots.$

In (i), we have a first-order Riccati $\Delta.E.$ as x_{n+1} depends on x_n. In (ii), we have a second-order Max-Type $\Delta.E.$ as x_{n+2} depends on x_{n+1} and x_n. Our aims are to solve difference equations by iterations (recursively) and inductively obtain an explicit solution.

Example 1.5 *Determine the* **explicit solution** *to the following $\Delta.E.$:*

$$x_{n+1} = ax_n, \quad n = 0, 1, \ldots,$$

Solution: *By iterations, we obtain the following pattern:*

$$x_0,$$
$$x_1 = ax_0,$$
$$x_2 = ax_1 = a \cdot [ax_0] = a^2 x_0,$$
$$x_3 = ax_2 = a \cdot [a^2 x_0] = a^3 x_0,$$
$$x_4 = ax_3 = a \cdot [a^3 x_0] = a^4 x_0,$$
$$\vdots$$

Note that $a \neq 0$ and for all $n \in \mathbb{N}$:

$$x_n = a^n x_0.$$

This is a solution to the first-order linear homogeneous $\Delta.E$.

1.3 Non-Autonomous Difference Equations

We will encounter several non-autonomous difference equations throughout the book: non-autonomous Linear Difference Equations, non-autonomous Riccati Difference Equations, non-autonomous Piece-wise Difference Equations and non-autonomous Max-Type Difference Equations. We will emanate with first-order non-autonomous linear difference equations.

(i) $x_{n+1} = a_n x_n, \quad n = 0, 1, \ldots,$

(ii) $x_{n+1} = x_n + b_n, \quad n = 0, 1, \ldots,$

(iii) $x_{n+1} = a_n x_n + b_n, \quad n = 0, 1, \ldots,$

where $\{a_n\}_{n=0}^{\infty}$ and $\{b_n\}_{n=0}^{\infty}$ are either sequences of real numbers or periodic sequences. In (i), each term of the sequence $\{a_n\}_{n=0}^{\infty}$ is multiplied by x_n during each iteration for all $n \geq 0$. In (ii) on the other hand, each term of the sequence $\{b_n\}_{n=0}^{\infty}$ is added to x_n during each iteration for all $n \geq 0$. In (iii), for all $n \geq 0$, b_n is added to the product of a_n and x_n. Therefore, there are three possibilities to express a first-order non-autonomous $\Delta.E$.

Example 1.6 *Determine an* **explicit solution** *to the following $\Delta.E$.:*

$$x_{n+1} = x_n + b_n, \quad n = 0, 1, \ldots,$$

where $\{b_n\}_{n=0}^{\infty}$ is a sequence of real numbers.

Solution: *Observe:*

$$\begin{aligned}
&x_0, \\
&x_1 = x_0 + b_0, \\
&x_2 = x_1 + b_1 = [x_0 + b_0] + b_1, \\
&x_3 = x_2 + b_2 = [x_0 + b_0 + b_1] + b_2, \\
&x_4 = x_3 + b_3 = [x_0 + b_0 + b_1 + b_2] + b_3,
\end{aligned}$$

$$\vdots$$

Hence for all $n \in \mathbb{N}$:

$$x_n = x_0 + \left[\sum_{i=0}^{n-1} b_i \right].$$

Example 1.7 *Determine an* **explicit solution** *to the following $\Delta.E.$:*

$$x_{n+1} = a_n x_n, \quad n = 0, 1, \dots,$$

where $\{a_n\}_{n=0}^{\infty}$ is a sequence of real numbers.
Solution: *Observe:*

$$x_0,$$
$$x_1 = a_0 x_0,$$
$$x_2 = a_1 x_1 = a_1 \cdot [a_0 x_0] = a_1 a_0 x_0,$$
$$x_3 = a_2 x_2 = a_2 \cdot [a_1 a_0 x_0] = a_2 a_1 a_0 x_0,$$
$$x_4 = a_3 x_3 = a_3 \cdot [a_2 a_1 a_0 x_0] = a_3 a_2 a_1 a_0 x_0,$$
$$\vdots$$

Hence for all $n \in \mathbb{N}$:

$$x_n = \left[\prod_{i=0}^{n-1} a_i \right] x_0.$$

1.4 Periodic Sequences

Our primary aim is to discern periodic traits and patterns graphically and analytically. Throughout the book, we will encounter assorted shapes of periodic cycles such as increasing period cycles, decreasing periodic cycles, alternating periodic cycles, oscillatory periodic cycles, triangular-shaped periodic cycles, trapezoidal-shaped periodic cycles and step-shaped periodic cycles. This section's goal is to introduce various periodic shapes with different periods.

In Algebra and Calculus we studied functions that rendered periodic behavior such as $y = \sin(x)$, $y = \cos(x)$, and piece-wise functions. The first graph (Figure 1.4) of a piece-wise function resembles a period-2 cycle:

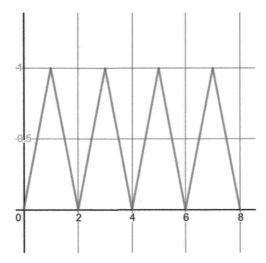

FIGURE 1.4: Piece-wise Function Describing a Period-2 Cycle.

In Figure 1.4, the piece-wise function is assembled with combinations of diagonal lines on restricted intervals with slopes 1 and -1. The next graph (Figure 1.5) outlines an **increasing** period-3 cycle:

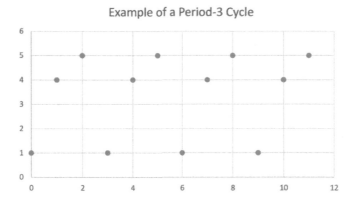

FIGURE 1.5: An Increasing Period-3 Cycle.

Observe that the terms of the period-3 cycle in Figure 1.5 increase from 1–4–5. The consequent sketch (Figure 1.6) evokes an **ascending triangular-shaped** period-3 cycle:

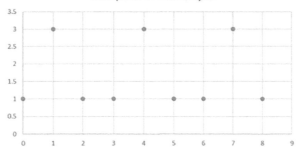

FIGURE 1.6: Triangular-Shaped Period-3 Cycle.

In Figure 1.6, the terms 1, 3, and 1 render the period-3 cycle in a triangular pattern. In Chapters 2 and 3, we will discover period-3 cycles in descending triangular-shaped patterns and in step-shaped patterns. Periodic cycles with period-4 and higher can also emerge in various shapes. The succeeding graph (Figure 1.7) traces an **ascending trapezoidal-shaped** period-4 cycle:

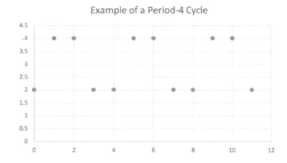

FIGURE 1.7: An Ascending Trapezoidal-Shaped Period-4 Cycle.

Now we will introduce several definitions.

Definition 1.1 *The sequence $\{x_n\}_{n=0}^{\infty}$ is periodic with* **minimal period-p,** *$(p \geq 2)$, provided that*

$$x_{n+p} = x_n \quad \text{for all} \quad n = 0, 1, \ldots.$$

Definition 1.2 *The sequence $\{x_n\}_{n=0}^{\infty}$ is an* **increasing sequence** *or an* **ascending sequence,** *provided that*

$$x_n \leq x_{n+1} \quad \text{for all} \quad n = 0, 1, \ldots.$$

Definition 1.3 *The sequence $\{x_n\}_{n=0}^{\infty}$ is a* **decreasing sequence** *or a* **descending sequence,** *provided that*

$$x_n \geq x_{n+1} \quad \text{for all} \quad n = 0, 1, \ldots.$$

The succeeding examples will render assorted periodic patterns analytically by solving the given **Initial Value Problem**. The first example describes an increasing period-3 cycle of a Piece-wise Linear Difference Equation (Tent-Map).

Example 1.8 *Solve the* **Initial Value Problem** *explicitly and determine the period of:*

$$
x_{n+1} = \begin{cases} 2x_n & \text{if } x_n < \frac{1}{2}, \\ 2(1 - x_n) & \text{if } x_n \geq \frac{1}{2} \\ x_0 = \frac{2}{9}, \end{cases} \quad n = 0, 1, \dots,
$$

Solution: *By iteration we obtain:*

$$
x_0 = \frac{2}{9},
$$

$$
x_1 = 2 \cdot x_0 = \frac{4}{9},
$$

$$
x_2 = 2 \cdot x_1 = \frac{8}{9},
$$

$$
x_3 = 2 (1 - x_2) = \frac{2}{9} = x_0.
$$

Thus we obtain an **increasing** *period-3 cycle resembled by the corresponding sketch (Figure 1.8):*

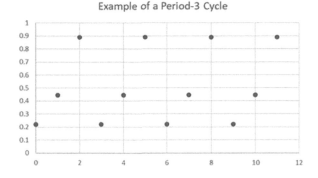

FIGURE 1.8: An Increasing Period-3 Cycle.

The next example renders **descending** and **ascending** period-4 cycles of a non-autonomous Riccati Difference Equation.

Example 1.9 *Solve the* **Initial Value Problem** *explicitly and determine the period of:*

$$
\begin{cases} x_{n+1} = \frac{(-1)^n}{x_n}, & n = 0, 1, \dots, \\ x_0 = 1. \end{cases}
$$

Solution: *By iteration we obtain:*

$$x_0 = 1,$$
$$x_1 = \frac{1}{x_0} = 1,$$
$$x_2 = \frac{-1}{x_1} = -1,$$
$$x_3 = \frac{1}{x_2} = -1,$$
$$x_4 = \frac{-1}{x_3} = 1.$$

This pattern traces a **descending step-shaped** *and an* **alternating** *period-4 cycle as* $x_0 = -x_2$ *and as* $x_1 = -x_3$ *depicted by the corresponding diagram (Figure 1.9):*

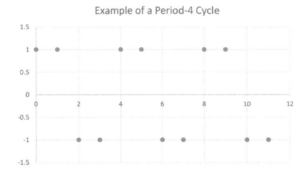

FIGURE 1.9: A Descending Step-Shaped and Alternating Period-4 Cycle.

Analogous to Figure 1.9, we can produce an **ascending step-shaped** *and an* **alternating** *period-4 cycle with* $x_0 = -1$ *(Figure 1.10):*

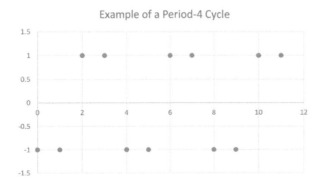

FIGURE 1.10: An Ascending Step-Shaped and Alternating Period-4 Cycle.

1.5   Alternating Periodic Cycles

Our aim is to detect distinct patterns of periodic cycles. How do we detect the pattern(s) when we transition from one neighbor of the periodic cycle to the next? This section will define and remit **alternating periodic sequences**.

Definition 1.4 $\{x_n\}_{n=0}^{\infty}$ *is an* **alternating periodic sequence** *with period-2k if for some* $k \in \mathbb{N}$, $x_n = -x_{n+k}$ *for all* $n \geq 0$.

Example 1.10 *The corresponding sketch (Figure 1.11):*

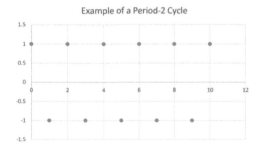

FIGURE 1.11: An Alternating Period-2 Cycle.

depicts an descending and an alternating period-2 cycle:

$$1, -1, 1, -1 \ \ldots,$$

as $x_n = -x_{n+1}$ *for all* $n \geq 0$.

Observe that period-2 cycles can emerge as either ascending period-2 patterns or as descending period-2 patterns.

Example 1.11 *The related graph (Figure 1.12):*

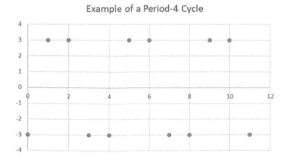

FIGURE 1.12: An Alternating Period-4 Cycle.

evokes an alternating period-4 cycle:

$$-3, 3, 3, -3, -3, 3, 3, -3, \ldots,$$

as $x_n = -x_{n+2}$ for all $n \geq 0$.

Notice that Figure 1.12 renders an ascending trapezoidal-shaped period-4 pattern. In contrast to Example (1.10), period-4 cycles can appear in various shapes; not necessarily as either ascending or descending patterns. We will see supplemental examples of period-4 cycles composed of different shapes in later chapters.

Example 1.12 *The consequent graph (Figure 1.13):*

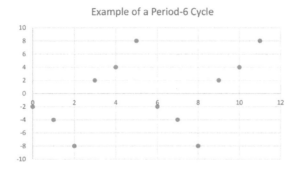

FIGURE 1.13: An Alternating Period-6 Cycle.

traces an alternating period-6 cycle as $x_n = -x_{n+3}$ for all $n \geq 0$.

1.6 Specific Patterns of Periodic Cycles

Our primary aim is to identify patterns of periodic cycles. In addition to ascending, descending and alternating patterns, how do we detect the pattern(s) when we transition from one neighbor of the cycle to the next? We will consider the periodic sequence $\{A_n\}_{n=0}^{\infty}$ with various periods. The next series of examples will explore the meticulous details of periodic structures together with the modulo arithmetic.

Example 1.13 *In the corresponding period-2 pattern:*

$$\frac{A_0}{1 - A_1}, \frac{A_1}{1 - A_0}, \ldots,$$

the indices of the period-2 sequence $\{A_n\}_{n=0}^{\infty}$ shift by 1 under modulo 2 arithmetic from neighbor to neighbor in the numerator and in the denominator.

Example 1.14 *In the associated period-3 pattern:*

$$\frac{A_0 - A_1 + A_2}{3}, \frac{A_1 - A_2 + A_0}{3}, \frac{A_2 - A_0 + A_1}{3}, \ldots,$$

the indices of the period-3 sequence $\{A_n\}_{n=0}^{\infty}$ shift by 1 under modulo 3 arithmetic from neighbor to neighbor in the numerator only while 3 (the term) in the denominator remains constant. In addition, in each term of the period-3 pattern, one component in the numerator is negative and two components are positive.

Example 1.15 *In the period-4 pattern:*

$$\frac{A_0 - A_1}{A_0 A_1 A_2 A_3}, \frac{A_1 - A_2}{A_0 A_1 A_2 A_3}, \frac{A_2 - A_3}{A_0 A_1 A_2 A_3}, \frac{A_3 - A_0}{A_0 A_1 A_2 A_3} \cdots,$$

the indices of the period-4 sequence $\{A_n\}_{n=0}^{\infty}$ shift by 1 under modulo 4 arithmetic from neighbor to neighbor in the numerator while the negative sign switches. Note that in this instance, all the terms in the period-4 cycle add up to 0.

Example 1.16 *In the period-4 pattern:*

$$\frac{A_0}{1 + A_0}, \frac{A_1}{1 + A_0 A_1}, \frac{A_2}{1 + A_0 A_1 A_2}, \frac{A_3}{1 + A_0 A_1 A_2 A_3}, \ldots,$$

the indices of the period-4 sequence $\{A_n\}_{n=0}^{\infty}$ shift by 1 under modulo 4 arithmetic from neighbor to neighbor in the numerator. In the denominator on the other hand, the terms of the sequence $\{A_n\}_{n=0}^{\infty}$ are multiplied.

More detailed examples of periodic patterns can be found in the **Appendix Chapter**.

1.7　Eventually Periodic Sequences

This section will focus on **eventually periodic sequences** and **eventually constant sequences**. We will commence with some definitions.

Definition 1.5 *The sequence $\{x_n\}_{n=0}^{\infty}$ is **Eventually Periodic** with* **minimal period (p $\geq$ 2)** *if there exists $N \geq 1$ such that*

$$x_{n+N+p} = x_{n+N} \quad \text{for all} \quad n = 0, 1, \ldots,$$

where N is the number of **transient terms**.

Definition 1.6 *The sequence $\{x_n\}_{n=0}^{\infty}$ is **Eventually Constant** if there exists $N \geq 1$ such that*

$$x_{n+N} = x_{n+N+1} = C \quad \text{for all} \quad n = 0, 1, \ldots,$$

where N is the number of **transient terms**.

The upcoming examples will depict eventually periodic solutions and eventually constant solutions together with transient terms (transient behavior).

Example 1.17 *The sequence below is eventually periodic with period-2 with five transient terms:*

$$[\mathbf{64}, \mathbf{32}, \mathbf{16}, \mathbf{8}, \mathbf{4}], 2, 1, 2, 1, \ldots.$$

The five transient terms $(x_0 - x_4)$ are in square brackets, $x_5 = x_7$ and $x_{n+5} = x_{n+7}$ for all $n \geq 0$. This is **3x+1 Conjecture** *that we will examine in Chapter 3.*

Now we will inspect some graphical examples of eventually periodic solutions and eventually constant solution with a different number of transient terms and their related assorted patterns.

Example 1.18 *The corresponding sketch (Figure 1.14) illustrates an eventually periodic cycle with period-2 with ten transient terms:*

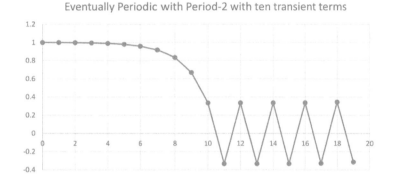

FIGURE 1.14: Eventually Periodic with ten Decreasing Transient Terms.

Figure 1.14 depicts an alternating period-2 cycle with ten descending transient terms $(x_0 - x_9)$. Also, $x_{10} = x_{12}$ and $x_{n+10} = x_{n+12}$ for all $n \geq 0$.

Example 1.19 *The related diagram (Figure 1.15) traces an eventually periodic cycle with period-4 with six transient terms:*

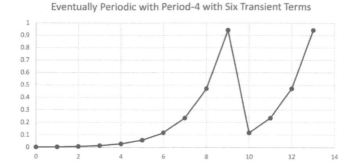

FIGURE 1.15: Eventually Periodic with Six Increasing Transient Terms.

Figure 1.15 depicts an ascending period-4 cycle with six ascending transient terms $(x_0 - x_5)$. In addition, $x_6 = x_{10}$ and $x_{n+6} = x_{n+10}$ for all $n \geq 0$.

Figures 1.14 and 1.15 render monotonically increasing and decreasing transient terms. The upcoming two examples will illustrate patterns of transient terms as decomposed piece-wise sequences.

Example 1.20 *The cognate graph (Figure 1.16) depicts an eventually periodic cycle with period-4 with eleven transient terms:*

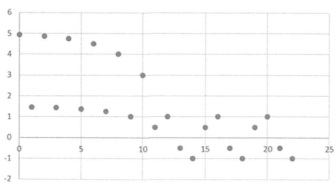

FIGURE 1.16: Eventually Periodic with Eleven Increasing Transient Terms.

Figure 1.16 depicts an eventually periodic cycle with period-4 with eleven transient terms $(x_0 - x_{10})$ as two descending piece-wise sequences. In addition, $x_{11} = x_{15}$ and $x_{n+11} = x_{n+15}$ for all $n \geq 0$.

Figures 1.14–1.16 are examples of **Piece-wise Difference Equations** that we will thoroughly examine in Chapter 4.

Example 1.21 *The corresponding graph (Figure 1.17) describes an eventually periodic cycle with period-2 with six transient terms:*

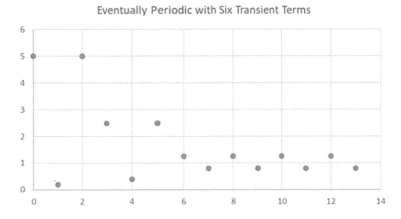

FIGURE 1.17: Eventually Periodic with Period-2 with Six Transient Terms.

Figure 1.17 portrays a descending period-2 cycle with six transient terms $(x_0 - x_5)$ *as three decomposed piece-wise sequences. In addition,* $x_6 = x_8$ *and* $x_{n+6} = x_{n+8}$ *for all* $n \geq 0$*. Figure 1.17 is an example of a* **Max-Type** $\Delta.E.$*, which we will explore in Chapter 5.*

Example 1.22 *The cognate sketch (Figure 1.18) renders a* **Piece-wise** $\Delta.E.$ *with an eventually constant sequence with five transient terms:*

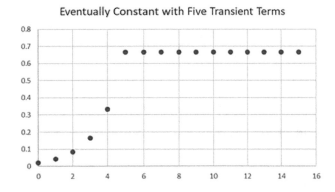

FIGURE 1.18: Eventually Constant Sequence with Five Transient Terms.

Figure 1.18 portrays an eventually constant sequence with five ascending transient terms $(x_0 - x_4)$*,* $x_5 = x_6$ *and* $x_{n+5} = x_{n+6}$ *for all* $n \geq 0$*.*

1.8   Piece-wise Sequences

This section's objective is to examine piece-wise sequences that consist of two or more sub-sequences.

Definition 1.7 $\{x_{n_k}\}_{k=0}^{\infty}$ is a **sub-sequence** of $\{x_n\}_{n=0}^{\infty}$.

We will commence with two corresponding examples of sub-sequences.

Example 1.23 $\{2n\}_{n=1}^{\infty}$ *is a linear* **sub-sequence** *of* $\{n\}_{n=1}^{\infty}$*:*

$$1, 2, 3, 4, 5, 6, 7, 8, 9, 10, \ldots.$$

Example 1.24 $\{4^n\}_{n=0}^{\infty}$ *is a geometric* **sub-sequence** *of* $\{2^n\}_{n=0}^{\infty}$*:*

$$1, 2, 4, 8, 16, 32, 64, 128, 256, 512, 1024, \ldots.$$

Definition 1.8 *For all* $n \geq 0$*, we define a* **Piece-wise Sequence** $\{x_n\}_{n=0}^{\infty}$ *that consists of two sub-sequences* $\{a_n\}_{n=0}^{\infty}$ *and* $\{b_n\}_{n=1}^{\infty}$ *as:*

$$a_0, b_1, a_2, b_3, \ldots. \tag{1.1}$$

Now we express (1.1) as:

$$\{x_n\}_{n=0}^{\infty} = \begin{cases} a_n & \text{if } n = 0, 2, 4, \ldots, \\ b_n & \text{if } n = 1, 3, 5, \ldots. \end{cases} \tag{1.2}$$

Example 1.25 *Write a formula for the following sequence:*

$$0, 1, 4, 5, 8, 9, 12, 13, \ldots. \tag{1.3}$$

Solution: *First we break up (1.3) into two main blue and green sub-sequences:*

$$0, 1, 4, 5, 8, 9, 12, 13, \ \ldots. \tag{1.4}$$

Then for $n \geq 0$*, we acquire:*

$$\{x_n\}_{n=0}^{\infty} = \begin{cases} 2n & \text{if } n = 0, 2, 4, 6, \ldots, \\ 2n - 1 & \text{if } n = 1, 3, 5, 7, \ldots. \end{cases}$$

Example 1.26 *Write a formula for the following sequence:*

$$1, C_0, C_0 C_1, C_0^2 C_1, C_0^2 C_1^2, C_0^3 C_1^2, C_0^3 C_1^3, C_0^4 C_1^3, \ldots. \tag{1.5}$$

Solution: *We will decompose (1.5) into two principal blue and green sub-sequences:*

$$1, C_0, [C_0 C_1], C_0 [C_0 C_1], [C_0 C_1]^2, C_0 [C_0 C_1]^2, [C_0 C_1]^3, C_0 [C_0 C_1]^3, \ \ldots. \tag{1.6}$$

Thus for $n \geq 0$*, we obtain:*

$$\{x_n\}_{n=0}^{\infty} = \begin{cases} [C_0 C_1]^{\frac{n}{2}} & \text{if } n = 0, 2, 4, 6, \ldots, \\ C_0 [C_0 C_1]^{\frac{n-1}{2}} & \text{if } n = 1, 3, 5, 7, \ldots. \end{cases}$$

Example 1.27 *Write a formula for the following sequence:*

$$2, 4, 6, -8, 10, 12, 14, -16, \ldots. \tag{1.7}$$

Solution: *Note that (1.7) is composed in terms of positive even integers starting at 2, while every fourth term of (1.7) is negative. Thus we will break up (1.7) into two primary blue and green sub-sequences:*

$$2, 4, 6, -8, 10, 12, 14, -16, \ldots \tag{1.8}$$

Now observe that in (1.8) the blue sub-sequence is a non-alternating sequence, while the green sub-sequence alternates. Hence for $n \geq 0$, we acquire:

$$\{x_n\}_{n=0}^{\infty} = \begin{cases} 2(n+1) & \text{if } n = 0, 2, 4, 6, \ldots, \\ (-1)^{\frac{n-1}{2}} [2(n+1)] & \text{if } n = 1, 3, 5, 7, \ldots. \end{cases}$$

Piece-wise sequences will be an essential tool in solving non-autonomous difference equations and describing the transient terms of eventually periodic solutions in Chapters 4 and 5. We will come upon the use of piece-wise sequences in Chapters 2–5. The upcoming two examples will apply piece-wise sequences in solving a specific initial value problem.

Example 1.28 *Solve the Initial Value Problem:*

$$\begin{cases} x_{n+1} = -x_n + (-1)^{n+1}, & n = 0, 1, \ldots, \\ x_0 = 4. \end{cases}$$

Solution: *By iterations we procure:*

$$\begin{aligned} x_0 &= 4, \\ x_1 &= -4 - 1 = -5, \\ x_2 &= 5 + 1 = 6, \\ x_3 &= -6 - 1 = -7, \\ x_4 &= 7 + 1 = 8, \\ x_5 &= -8 - 1 = -9, \\ &\vdots \end{aligned} \tag{1.9}$$

Thus via (1.9) we procure the cognate piece-wise sequence:

$$\{x_n\}_{n=0}^{\infty} = \begin{cases} [n+4] & \text{if } n = 0, 2, 4, 6, \ldots, \\ -[n+4] & \text{if } n = 1, 3, 5, 7, \ldots. \end{cases}$$

Example 1.29 *Solve the Initial Value Problem:*

$$\begin{cases} x_{n+1} = x_n + b_n, & n = 0, 1, \ldots, \\ x_0 = 0, \end{cases}$$

where:

$$\{b_n\}_{n=0}^{\infty} = \begin{cases} b_0 & \text{if } n = 0, 2, 4, 6, \ldots, \\ b_1 & \text{if } n = 1, 3, 5, 7, \ldots. \end{cases}$$

Solution: *By iterations we obtain:*

$$
\begin{aligned}
x_0 &= 0, \\
x_1 &= b_0, \\
x_2 &= b_0 + b_1, \\
x_3 &= 2b_0 + b_1, \\
x_4 &= 2b_0 + 2b_1, \\
x_5 &= 3b_0 + 2b_1,
\end{aligned}
\tag{1.10}
$$

$$
\vdots
$$

Via (1.10) we acquire the associated piece-wise sequence:

$$
\{x_n\}_{n=0}^{\infty} = \begin{cases} \frac{n[b_0 + b_1]}{2} & \text{if } n = 0, 2, 4, 6, \ldots, \\ \frac{(n+1)}{2} b_0 + \frac{(n-1)}{2} b_1 & \text{if } n = 1, 3, 5, 7, \ldots. \end{cases}
$$

1.9   Chapter 1 Exercises

In problems 1–8, write a **recursive formula** (as an initial value problem) of each sequence:

1: $5, 10, 15, 20, 25, 30, 35, \ldots.$

2: $7, 11, 15, 19, 23, 27, 31, \ldots.$

3: $3, 5, 9, 15, 23, 33, 45, \ldots.$

4: $5, 8, 14, 23, 35, 50, 68, \ldots.$

5: $1, 5, 13, 25, 41, 61, 85, \ldots.$

6: $2, 7, 12, 17, 22, 27, 32, \ldots.$

7: $1, 4, 13, 28, 49, 76, 109, \ldots.$

8: $6, 8, 14, 24, 38, 56, 78, \ldots.$

In problems 9–18, write a **recursive formula** (as an initial value problem) of each sequence:

9: $3, 12, 48, 192, 768, 3072, \ldots.$

10: $5, 10, 20, 40, 80, 160, \ldots.$

11: $54, 36, 24, 16, \frac{32}{3}, \frac{64}{9}, \ldots.$

12: $32, 24, 18, \frac{27}{2}, \frac{81}{8}, \frac{343}{32}, \ldots.$

13: $1, 3, 18, 162, 1944, 29160, \ldots.$

14: $2, 4, 24, 240, 3360, 60480, \ldots.$

15: $1, 5, 45, 585, 9945, 208845 \ldots$.

16: $1 \cdot 2, 2 \cdot 3, 3 \cdot 4, 4 \cdot 5, 5 \cdot 6, \ldots$.

17: $1 \cdot 3, 3 \cdot 5, 5 \cdot 7, 7 \cdot 9, 9 \cdot 11, \ldots$.

18: $2 \cdot 4, 4 \cdot 6, 6 \cdot 8, 8 \cdot 10, 10 \cdot 12, \ldots$.

In problems 19–22, write a **recursive formula** (as an initial value problem) of each summation:

19: $\sum_{k=1}^{n} k^2$.

20: $\sum_{k=0}^{n} (3k + 2)$.

21: $\sum_{k=1}^{n} (2k - 1)$.

22: $\sum_{k=0}^{n} \left(\frac{1}{2}\right)^k$.

In problems 23–30, determine the **period** of each Initial Value Problem:

23: $\begin{cases} x_{n+1} = \frac{6}{x_n}, n = 0, 1, \ldots, \\ x_0 = \frac{1}{2}. \end{cases}$

24: $\begin{cases} x_{n+1} = \frac{1+x_n}{x_n-1}, n = 0, 1, \ldots, \\ x_0 = 3. \end{cases}$

25: $\begin{cases} x_{n+1} = x_n + (-1)^n, n = 0, 1, \ldots, \\ x_0 = 4. \end{cases}$

26: $\begin{cases} x_{n+1} = (-1)^n x_n - 1, n = 0, 1, \ldots, \\ x_0 = -3. \end{cases}$

27: $\begin{cases} x_{n+1} = \frac{(-1)^{n+1}}{x_n}, n = 0, 1, \ldots, \\ x_0 = -4. \end{cases}$

28: $\begin{cases} x_{n+2} = \frac{8}{x_n}, n = 0, 1, \ldots, \\ x_0 = 2, \\ x_1 = 4. \end{cases}$

29: $\begin{cases} x_{n+2} = \frac{(-1)^n}{x_n}, n = 0, 1, \ldots, \\ x_0 = 1, \\ x_1 = 2. \end{cases}$

30: $\begin{cases} x_{n+2} = \frac{x_{n+1}}{x_n}, n = 0, 1, \ldots, \\ x_0 = 2, \\ x_1 = 6. \end{cases}$

In problems 31–36, determine the period and structure of the periodic cycles from the corresponding graphs:

31. Period of:

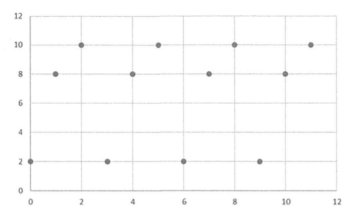

32. Period of:

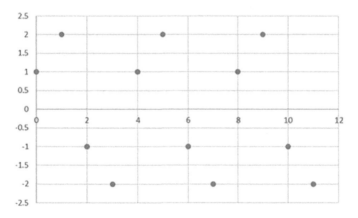

33. Period of:

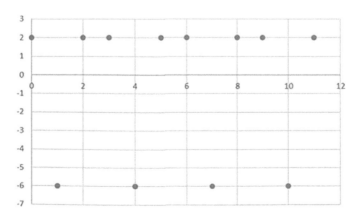

34. Period of:

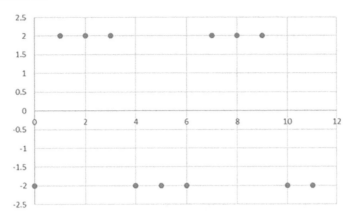

35. Period of:

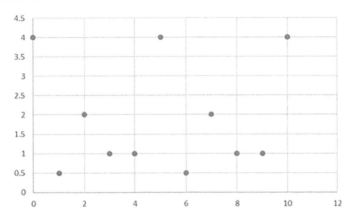

36. Period of:

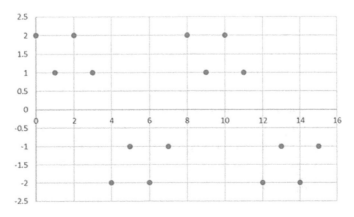

In problems 37–48, write a **formula** of each piece-wise sequence:

37: $1, 5, 6, 10, 11, 15, 16, 20, \ldots$

38: $2, 4, 9, 11, 16, 18, 23, 25, \ldots$

39: $3, 6, 12, 15, 21, 24, 30, 33, \ldots$

40: $4, 8, 16, 20, 28, 32, 40, 44, \ldots$

41: $1, 3, 3, 9, 9, 27, 27, 81, 81, \ldots$

42: $1, 2, 8, 16, 64, 128, 512, 1024, \ldots$

43: $2, 6, 12, 36, 72, 216, 432, 1296, \ldots$

44: $4, 20, 40, 200, 400, 2000, 4000, 20000, \ldots$

45: $1, -3, 6, -12, 18, -24, 30, -36, \ldots$

46: $-2, 8, -14, 20, -26, 32, -38, 44, \ldots$

47: $2, 4, 6, -8, 10, 12, 14, -16, \ldots$

48: $-4, 8, 12, 16, -20, 24, 28, 32, \ldots$

In problems 49–54, as in Examples (1.13)–(1.16), describe the pattern of the given periodic cycle:

49. $\dfrac{1 - a_1}{1 - a_0 a_1}, \ \dfrac{1 - a_0}{1 - a_0 a_1}, \ldots$

50. $\sqrt{\dfrac{A_0 A_1}{A_2}}, \ \sqrt{\dfrac{A_1 A_2}{A_0}}, \ \sqrt{\dfrac{A_2 A_0}{A_1}}, \ldots$

51. $\dfrac{b_0 + b_1 + b_2 - b_3}{1 - b_0 b_1 b_2 b_3}, \ \dfrac{b_0 + b_1 - b_2 + b_3}{1 - b_0 b_1 b_2 b_3}, \ \dfrac{b_0 - b_1 + b_2 + b_3}{1 - b_0 b_1 b_2 b_3},$

$\dfrac{-b_0 + b_1 + b_2 + b_3}{1 - b_0 b_1 b_2 b_3}, \ldots$

52. $\dfrac{a_0}{1 + a_0}, \ \dfrac{a_0 a_1}{1 + a_0 + a_1}, \ \dfrac{a_0 a_1 a_2}{1 + a_0 + a_1 + a_2}, \ldots$

53. $\dfrac{a_1 b_3 - b_1}{a_1 a_3 + 1}, \ \dfrac{a_2 b_0 - b_2}{a_0 a_2 + 1}, \ \dfrac{a_3 b_1 - b_3}{a_1 a_3 + 1}, \ \dfrac{a_0 b_2 - b_0}{a_0 a_2 + 1}, \ldots$

54. $\dfrac{b_0 b_1 b_2 b_3 + 1}{b_0 b_1 b_2 - b_0 b_1 + b_0 - 1}, \ \dfrac{b_0 b_1 b_2 b_3 + 1}{b_1 b_2 b_3 - b_1 b_2 + b_1 - 1}, \ \dfrac{b_0 b_1 b_2 b_3 + 1}{b_2 b_3 b_0 - b_2 b_3 + b_2 - 1},$

$\dfrac{b_0 b_1 b_2 b_3 + 1}{b_3 b_0 b_1 - b_3 b_0 + b_3 - 1}, \ldots$

Chapter 2

Linear Difference Equations

This chapter's aim is to introduce various periodic traits of the Autonomous and the Non-Autonomous First-Order Linear Difference Equations. We will encounter assorted periods with mixed shapes (ascending, descending, oscillatory, alternating, step-shaped, triangular-shaped and trapezoidal-shaped) and the existence of unique periodic cycles in some instances. We will start off with two graphical examples of period-3 patterns. For instance, the following diagram (Figure 2.1) renders an **ascending step-shaped** period-3 cycle:

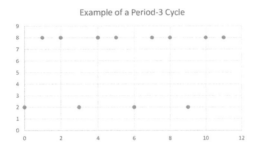

FIGURE 2.1: An Ascending Step-Shaped Period-3 Cycle.

Analogous graphs can be constructed that describe a descending step-shaped period-3 cycle; we will see such examples in a later sub-section that examines periodic traits of Non-Autonomous Linear Difference Equations. The consequent graph (Figure 2.2) evokes an **ascending triangular-shaped** period-3 cycle:

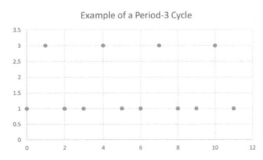

FIGURE 2.2: An Ascending Triangular-Shaped Period-3 Cycle.

We will now commence our study with periodic traits of first-order Linear Autonomous Difference Equations.

2.1 Autonomous Linear Difference Equations

We will commence our study of periodic properties of the corresponding first-order autonomous linear difference equation in the form:

$$x_{n+1} = -x_n + b, \quad n = 0, 1, \ldots, \tag{2.1}$$

where $x_0, b \in \Re$. Eq. (2.1) consists of two cases when $b = 0$ and $b \neq 0$. When $b = 0$, Eq. (2.1) reduces to the following homogeneous linear difference equation:

$$x_{n+1} = -x_n, \quad n = 0, 1, \ldots, \tag{2.2}$$

where $x_0 \neq 0$. By iterations and induction we obtain the following solution to Eq. (2.2):

$$x_n = (-1)^n x_0 = \begin{cases} x_0 & \text{if n is even,} \\ -x_0 & \text{if n is odd.} \end{cases} \tag{2.3}$$

Figure 1.11 in Chapter 1 is a special case of (2.3) when $x_0 = 1$. The consequent sketch (Figure 2.3) portrays (2.3) as an **ascending alternating** period-2 cycle when $x_0 = -4$:

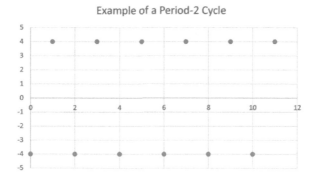

FIGURE 2.3: An Ascending Alternating Period-2 Cycle.

Now suppose that $b \neq 0$. Then the general solution of Eq. (2.1) is:

$$x_n = \begin{cases} x_0 & \text{if n is even,} \\ -x_0 + b & \text{if n is odd.} \end{cases} \tag{2.4}$$

We will now transition to periodic traits of non-autonomous linear difference equations.

2.2 Non-Autonomous Linear $\Delta.E.$'s

We will advance with the study of periodic traits of first-order non-autonomous linear $\Delta.E.$ in the form:

$$x_{n+1} = a_n x_n + b_n, \quad n = 0, 1, \ldots, \tag{2.5}$$

where $\{a_n\}_{n=0}^{\infty}$ and $\{b_n\}_{n=0}^{\infty}$ are periodic sequences with either the same period or different periods. We will examine Eq. (2.5) in the multiplicative and additive forms.

2.2.1 Multiplicative Form of Eq. (2.5)

We will commence with the first special case of Eq. (2.5) which is the following difference equation:

$$x_{n+1} = a_n x_n, \quad n = 0, 1, \ldots, \tag{2.6}$$

where $x_0 \neq 0$ and $\{a_n\}_{n=0}^{\infty}$ is a period-k sequence for $(k \geq 2)$. Our goal is to examine the periodic traits of Eq. (2.6). The first example renders the periodicity of Eq. (2.6) when $\{a_n\}_{n=0}^{\infty}$ is a period-2 sequence.

Example 2.1 *Solve Eq. (2.6) when $\{a_n\}_{n=0}^{\infty}$ is a period-2 sequence and determine the periodic traits.*

Solution: *By iterations and induction we get:*

$$x_0,$$
$$x_1 = a_0 x_0,$$
$$x_2 = a_1 x_1 = [a_1 a_0] x_0,$$
$$x_3 = a_0 x_2 = a_1 a_0^2 x_0,$$
$$x_4 = a_1 x_3 = [a_1 a_0]^2 x_0,$$
$$x_5 = a_0 x_4 = a_1^2 a_0^3 x_0,$$
$$x_6 = a_1 x_5 = [a_1 a_0]^3 x_0,$$
$$\vdots$$

Let $P = a_0 a_1$. Then for all $n \geq 0$:

$$\begin{cases} x_{2n} = P^n x_0, \\ x_{2n+1} = a_0 P^n x_0. \end{cases} \tag{2.7}$$

Note that via (2.11), every solution of Eq. (2.6) is periodic with:

(i) Period-2 if $P = 1$.

(ii) Period-4 if $P = -1$.

*Now we will portray graphical representations of a period-2 when $P = 1$ and a period-4 cycle when $P = -1$ and compare the patterns' disparities. The first graph (Figure 2.4) depicts a **positive** period-2 cycle when $x_0 = 2$, $a_0 = 4$ and $a_1 = 0.25$:*

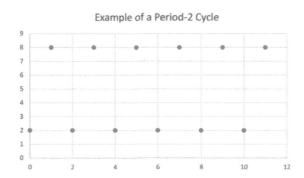

FIGURE 2.4: A Positive Period-2 Cycle.

*By designating distinct values of x_0, b_0 and b_1, we can acquire either a negative periodic cycle or a cycle with a positive and a negative term; this will be left as an end-of-chapter exercise. The consequent graph (Figure 2.5) renders an **ascending trapezoidal-shaped** and an **alternating** period-4 cycle when $x_0 = -2$, $a_0 = -1$ and $a_1 = 1$.*

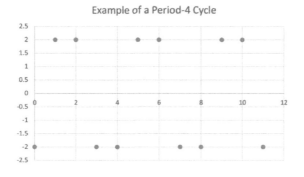

FIGURE 2.5: A Trapezoidal-Shaped Alternating Period-4 Cycle.

Figure 2.5 evokes an alternating period-4 cycle as $x_{n+2} = -x_n$ for all $n \geq 0$. When $a_0 = 1$ and $a_1 = -1$ or vice versa, Eq. (2.6) depicts **trapezoidal-shaped** *period-4 cycles and* **step-shaped** *period-4 cycles. Furthermore, if $P = -1$, then every solution of Eq. (2.6) is the corresponding alternating period-4 cycle:*

$$x_0, \ a_0 x_0, \ -x_0, \ -a_0 x_0, \ \ldots.$$

If on the other hand $a_0 \neq \pm 1$ and $a_1 \neq \pm 1$, then Eq. (2.6) will portray **alternating** *period-4 cycles with positive increasing terms and negative decreasing terms or vice versa. The sequential diagram (Figure 2.6) depicts an alternating period-4 cycle with two increasing terms and two decreasing terms when $x_0 = 4$, $a_0 = 2$ and $a_1 = -0.5$:*

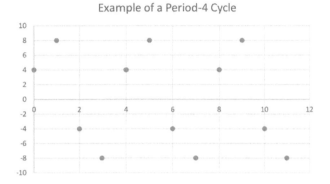

FIGURE 2.6: An Alternating Period-4 Cycle.

The sequential example will outline the periodic nature of Eq. (2.6) when $\{a_n\}_{n=0}^{\infty}$ is a period-3 sequence.

Example 2.2 *Solve Eq. (2.6) when $\{a_n\}_{n=0}^{\infty}$ is a period-3 sequence and determine the periodic traits.*

Solution: *Let $P = a_0 a_1 a_2$, then analogous to Example (2.1), for all $n \geq 0$ we acquire:*

$$
\begin{cases}
x_{3n} = P^n x_0, \\
x_{3n+1} = a_0 P^n x_0, \\
x_{3n+2} = a_0 a_1 P^n x_0.
\end{cases}
\tag{2.8}
$$

Thus via (2.12), every solution of Eq. (2.6) is periodic with:

(i) Period-3 if $P = 1$.

(ii) Period-6 if $P = -1$.

Now we will illustrate graphs of a period-3 when $P = 1$ and a period-6 cycle when $P = -1$ and compare the patterns' contrasts. In fact, we will discover the same differences analogous to Example (2.1): a **positive** *period-3 cycle and an* **alternating** *period-6 cycle. The first graph (Figure 2.7) depicts an* **ascending step-shaped** *period-3 cycle when $x_0 = 2$, $a_0 = 4$, $a_1 = 1$ and $a_2 = 0.25$:*

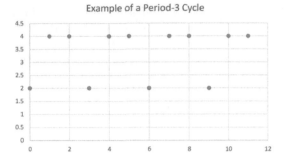

FIGURE 2.7: Ascending Step-Shaped Period-3 Cycle.

*Similar to Example (2.1), we can select values of x_0, a_0, a_1 and a_2 and obtain different periodic structures. For instance, the consequent diagram (Figure 2.8) traces a **decreasing** period-3 cycle when $x_0 = 2$, $a_0 = -2$, $a_1 = 2$ and $a_2 = -0.25$:*

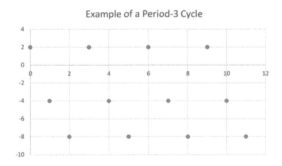

FIGURE 2.8: A Decreasing Period-3 Cycle with One Positive Term and Two Negative Terms.

*The next graph (Figure 2.9) renders a **step-shaped** and an **alternating** period-6 cycle when $x_0 = 2$, $a_0 = 2$, $a_1 = 1$ and $a_2 = -0.5$:*

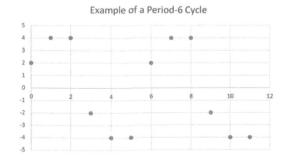

FIGURE 2.9: A Step-Shaped Alternating Period-6 Cycle.

The following sketch (Figure 2.10) describes a **step-shaped** *and an* **alternating** *period-8 cycle when* $x_0 = 2$, $a_0 = 2$, $a_1 = 1$, $a_2 = 2$ *and* $a_3 = -0.25$:

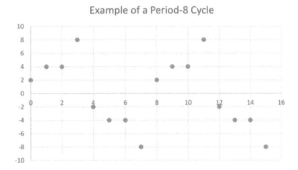

FIGURE 2.10: A Step-Shaped Alternating Period-8 Cycle.

Now suppose that $\{a_n\}_{n=0}^{\infty}$ is a period-k sequence ($k \geq 2$) and let $P = \prod_{i=0}^{k-1} a_i$. Then parallel to Examples (2.1) and (2.2) we acquire the following solution to Eq. (2.6):

$$\begin{cases} x_{kn} = P^n x_0, \\ x_{kn+1} = a_0 P^n x_0, \\ x_{kn+2} = a_0 a_1 P^n x_0, \\ \vdots \\ x_{kn+k-2} = \left[\prod_{i=0}^{k-3} a_i\right] P^n x_0, \\ x_{kn+k-1} = \left[\prod_{i=0}^{k-2} a_i\right] P^n x_0. \end{cases} \qquad (2.9)$$

Hence via (2.9), every solution of Eq. (2.6) is periodic with:

(i) Period-k if $P = 1$.

(ii) Period-2k if $P = -1$.

Via Examples (2.1) and (2.2), when $P = -1$, every solution of Eq. (2.6) is an **alternating** period-2k cycle, where for all $n \geq 0$, $x_{n+k} = -x_n$.

2.2.2 Additive Form of Eq. (2.5)

We will progress with periodic traits of the next special case of Eq. (2.5):

$$x_{n+1} = x_n + b_n, \quad n = 0, 1, \ldots, \qquad (2.10)$$

where $\{b_n\}_{n=0}^{\infty}$ is a period-k sequence ($k \geq 2$). Our goal is to determine the existence of periodic cycles of Eq. (2.10). The first example will examine the periodicity of Eq. (2.10) when $\{b_n\}_{n=0}^{\infty}$ as a period-2 sequence.

Example 2.3 *Solve Eq. (2.10) when* $\{b_n\}_{n=0}^{\infty}$ *is a period-2 sequence and determine the periodic character.*

Solution: *Analogous to Example (2.1), by iterations we acquire:*

$$x_0,$$
$$x_1 = x_0 + b_0,$$
$$x_2 = x_1 + b_1 = x_0 + [b_0 + b_1],$$
$$x_3 = x_2 + b_0 = x_0 + b_0 + [b_0 + b_1],$$
$$x_4 = x_3 + b_1 = x_0 + 2[b_0 + b_1],$$
$$x_5 = x_4 + b_0 = x_0 + b_0 + 2[b_0 + b_1],$$
$$x_6 = x_5 + b_1 = x_0 + 3[b_0 + b_1],$$
$$\vdots$$

Let $S = b_0 + b_1$. *Thus for all* $n \geq 0$:

$$\begin{cases} x_{2n} = x_0 + Sn, \\ x_{2n+1} = x_0 + b_0 + Sn. \end{cases} \tag{2.11}$$

The consequent graph (Figure 2.11) sketches a **positive increasing** *period-2 cycle when* $x_0 = 2$, $b_0 = 4$ *and* $b_1 = -4$:

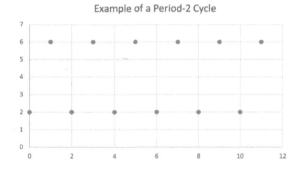

Example of a Period-2 Cycle

FIGURE 2.11: A Positive Period-2 Cycle.

Notice that via (2.11), every solution of Eq. (2.10) is periodic with period-2 if $S = 0$ *with the following pattern:*

$$x_0, \ x_0 + b_0, \ x_0, \ x_0 + b_0, \ \ldots.$$

By selecting different values of x_0, b_0 *and* b_1, *we can obtain either a negative periodic cycle or a periodic cycle with a positive and a negative term.*

Example 2.4 *Solve Eq. (2.10) when $\{b_n\}_{n=0}^{\infty}$ is a period-3 sequence and determine the periodic character.*

Solution: *Let $S = b_0 + b_1 + b_2$. Then analogous to Example (2.3), for all $n \geq 0$ we acquire:*

$$\begin{cases} x_{3n} = x_0 + Sn, \\ x_{3n+1} = x_0 + b_0 + Sn, \\ x_{3n+2} = x_0 + b_0 + b_1 + Sn. \end{cases} \tag{2.12}$$

Via (2.12), every solution of Eq. (2.10) is periodic with period-3 if $S = 0$. The corresponding graph (Figure 2.12) renders an **increasing** *period-3 cycle with two negative terms and one positive term when $x_0 = -4$, $b_0 = 2$, $b_1 = 3$ and $b_2 = -5$:*

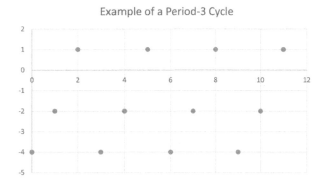

FIGURE 2.12: A Increasing Period-3 Cycle.

We can also detect the structure's sensitivity of the periodic cycles depending on the values of x_0, b_0, b_1 and b_2. Switching the negative sign(s) of either b_0, b_1 or b_2 will affect the pattern of the periodic cycle. Sometimes a periodic cycle will consist of only positive values, negative values, may be a decreasing periodic cycle or maybe neither increasing nor decreasing. The patterns' structures are very sensitive to what specific value or values of b_0, b_1 or b_2 are positive or negative. Supplemental problems will be provided at the end-of-chapter exercises.

Now we will portray additional examples of periodic traits of Eq. (2.10) when $\{b_n\}_{n=0}^{\infty}$ is a period-4 sequence and a period-6 sequence. Suppose that $\{a_n\}_{n=0}^{\infty}$ is a period-4 sequence. The sketch below (Figure 2.13) outlines a **decreasing** period-4 cycle with two positive terms, one zero term and one negative term when $x_0 = 6$, $b_0 = -2$, $b_1 = -4$, $b_2 = -6$ and $b_3 = 12$:

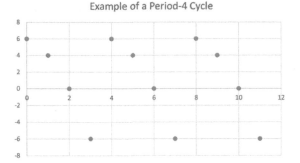

FIGURE 2.13: A Decreasing Period-4 Cycle.

Now assume that $\{b_n\}_{n=0}^{\infty}$ is a period-6 sequence. The graph below (Figure 2.14) describes an **oscillatory** period-6 cycle with one positive term and five negative terms when $x_0 = -1$, $b_0 = 2$, $b_1 = -3$, $b_2 = -2$, $b_3 = 1$, $b_4 = 1$ and $b_5 = 1$:

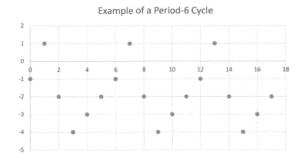

FIGURE 2.14: An Oscillatory Period-6 Cycle.

Suppose that $\{b_n\}_{n=0}^{\infty}$ is a period-k sequence ($k \geq 2$) and let $S = \sum_{i=0}^{k-1} b_i$. Then via Examples (2.3) and (2.4), we procure the following solution to Eq. (2.10):

$$\begin{cases} x_{kn} = x_0 + Sn, \\ x_{kn+1} = x_0 + b_0 + Sn, \\ x_{kn+2} = x_0 + b_0 + b_1 + Sn, \\ \vdots \\ x_{kn+k-2} = x_0 + \left[\sum_{i=0}^{k-3} b_i\right] + Sn, \\ x_{kn+k-1} = x_0 + \left[\sum_{i=0}^{k-2} b_i\right] + Sn. \end{cases} \qquad (2.13)$$

It follows via (2.13) that every solution of Eq. (2.10) is periodic with period-k if $S = 0$.

We will next shift our focus to the periodic features of the following special case of Eq. (2.5):

$$x_{n+1} = -x_n + b_n, \quad n = 0, 1, \ldots, \tag{2.14}$$

where $\{b_n\}_{n=0}^{\infty}$ is a period-k sequence ($k \geq 2$). We will investigate the existence and uniqueness of periodic cycles of Eq. (2.14). The succeeding example will emerge with $\{b_n\}_{n=0}^{\infty}$ as a period-2 sequence.

Example 2.5 *Suppose that $\{b_n\}_{n=0}^{\infty}$ is a period-2 sequence. Show that Eq. (2.14) has no period-2 cycles and explain why.*

Solution: *Suppose that $x_2 = x_0$. Then we acquire:*

$$x_0,$$
$$x_1 = -x_0 + b_0,$$
$$x_2 = -x_1 + b_1 = -[x_0 + b_0] + b_1 = x_0 - b_0 + b_1 = x_0.$$

Thus $x_2 = x_0$ if and only if $b_0 = b_1$. This is a contradiction as we assumed that $\{b_n\}_{n=0}^{\infty}$ is a period-2 sequence where $b_0 \neq b_1$.

The next sequence of examples will signify the contrasting periodic traits of Eq. (2.14) when $\{b_n\}_{n=0}^{\infty}$ is an even-ordered periodic sequence in comparison to when $\{b_n\}_{n=0}^{\infty}$ is an odd-ordered periodic sequence.

Example 2.6 *Suppose that $\{b_n\}_{n=0}^{\infty}$ is a period-3 sequence. Determine the necessary and sufficient conditions for the existence of period-3 cycles of Eq. (2.14).*

Solution: *By iterations we acquire:*

$$x_0,$$
$$x_1 = -x_0 + b_0,$$
$$x_2 = -[x_1] + b_1 = -[-x_0 + b_0] + b_1 = x_0 + b_1 - b_0,$$
$$x_3 = -[x_2] + b_2 = -[x_0 + b_1 - b_0] + b_2 = -x_0 + b_2 + b_0 - b_1 = x_0.$$

Hence we obtain

$$x_0 = \frac{b_0 - b_1 + b_2}{2}, \tag{2.15}$$

and the corresponding **unique** *period-3 pattern:*

$$x_0 = \frac{b_0 - b_1 + b_2}{2}, \ x_1 = \frac{b_1 - b_2 + b_0}{2}, \ x_2 = \frac{b_2 - b_0 + b_1}{2}, \ \ldots \tag{2.16}$$

This is the also first time we encounter a **unique periodic cycle**. *Observe that the indices of the sequence $\{b_n\}_{n=0}^{\infty}$ in (2.16) shift forward by an index of 1 from term to term. The consequent sketch (Figure 2.15) renders (2.16) as a* **descending step-shaped** *period-3 cycle when $x_0 = 2$, $b_0 = 4$, $b_1 = 1$ and $b_2 = 1$:*

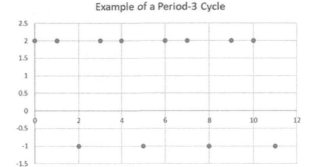

FIGURE 2.15: A Descending Step-Shaped Period-3 Cycle.

In all other cases, every solution of Eq. (2.14) is periodic with period-6; this will left as an end-of-chapter exercise to verify. The upcoming diagram (Figure 2.16) evokes a period-6 cycle of Eq. (2.14) when $x_0 = 1$, $b_0 = 4$, $b_1 = 1$ and $b_2 = 1$:

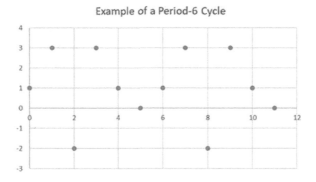

FIGURE 2.16: A Scattered Period-6 Cycle.

Example (2.6) then leads us to the following result, describing the unique periodic character of Eq. (2.14) when $\{b_n\}_{n=0}^{\infty}$ is an odd-ordered periodic sequence.

Theorem 2.1 *Suppose that $\{b_n\}_{n=0}^{\infty}$ is a period-$(2k+1)$ sequence, $(k \in \mathbb{N})$. Then one of the following statements holds true:*

*(1) Eq. (2.14) has a **unique periodic cycle** with period-$(2k+1)$ where*

$$x_0 = \frac{\sum_{i=0}^{k+1} b_{2i} - \sum_{i=0}^{k} b_{2i+1}}{2} = \frac{\sum_{i=1}^{2k+1} (-1)^{i+1} b_{i-1}}{2}, \qquad (2.17)$$

(2) Every solution of Eq. (2.14) is periodic with period-2 $(2k+1)$.

First of all, observe that (2.17) is an extension of (2.15). Second of all, notice that via (2.15) we can conclude that (2.17) has $k + 1$ positive terms and k negative terms. In fact, the even-indexed coefficients are positive and the odd-indexed coefficients are negative. The proof of (2.17) and Theorem (2.1) will be left as an end-of-chapter exercise. Now we will direct our focus to when $\{b_n\}_{n=0}^{\infty}$ is an even-ordered periodic sequence.

Example 2.7 *Suppose that $\{b_n\}_{n=0}^{\infty}$ is a period-4 sequence. Determine the necessary and sufficient conditions for the existence of period-4 cycles of Eq. (2.14).*

Solution: *By iteration, we procure:*

$$x_0,$$
$$x_1 = -x_0 + b_0,$$
$$x_2 = -[x_1] + b_1 \;=\; -[-x_0 + b_0] + b_1 \;=\; x_0 + b_1 - b_0,$$
$$x_3 = -[x_2] + b_2 \;=\; -[x_0 + b_1 - b_0] + b_2 \;=\; -x_0 + a_2 + a_0 - a_1$$
$$x_4 = -[x_3] + b_3 \;=\; -[-x_0 + b_2 + b_0 - b_1] + b_3,$$
$$= x_0 - [b_2 + b_0] + [b_1 + b_3] \;=\; x_0.$$

Period-4 cycles exist if and only if

$$b_1 + b_3 = b_0 + b_2 \tag{2.18}$$

with the corresponding period-4 pattern:

$$x_0, \;\; -x_0 + b_0, \;\; x_0 + b_1 - b_0, \;\; -x_0 + b_2 + b_0 - b_1, \;\; \ldots \tag{2.19}$$

The ensuing graph (Figure 2.17) renders (2.19) as a period-4 cycle with one positive term and three negative terms when $x_0 = 8$, $b_0 = 4$, $b_1 = -6$, $b_2 = -4$ and $b_3 = 6$:

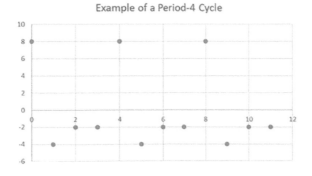

FIGURE 2.17: A Period-4 Cycle.

Example (2.7) then guides us to the upcoming result that portrays the periodic traits of Eq. (2.14) when $\{b_n\}_{n=0}^{\infty}$ is an even-ordered periodic sequence.

Theorem 2.2 *Suppose that $\{b_n\}_{n=0}^{\infty}$ is a period-2k sequence, $(k \geq 2)$. Then every solution of Eq. (2.14) is periodic with period-2k if and only if:*

$$\sum_{i=1}^{k} b_{2i-1} = \sum_{i=1}^{k} b_{2i-2}. \tag{2.20}$$

Notice that (2.20) extends directly from (2.18). All the odd-indexed terms of $\{b_n\}_{n=0}^{\infty}$ are grouped on the left side of (2.20) while all the even-indexed terms of $\{b_n\}_{n=0}^{\infty}$ are grouped on the right side of (2.20). In addition, the sum of all the odd-indexed coefficients must be equal to the sum of all the even-indexed coefficients. Furthermore, (2.20) has k terms on the left side and k terms on the right side. The proof of (2.20) will be left as an end-of-chapter exercise.

We will now turn our focus to the next special case of Eq. (2.5) in the additive and in the multiplicative form:

$$x_{n+1} = -a_n x_n + 1, \quad n = 0, 1, \ldots, \tag{2.21}$$

where $\{a_n\}_{n=0}^{\infty}$ is a period-k sequence $(k \geq 2)$. We will come across assorted periodic patterns when $\{a_n\}_{n=0}^{\infty}$ is an even-ordered and when $\{a_n\}_{n=0}^{\infty}$ is an odd-ordered periodic sequence. The upcoming examples will emphasize these contrasts when $\{a_n\}_{n=0}^{\infty}$ is a period-2 sequence and $\{a_n\}_{n=0}^{\infty}$ is a period-3 sequence.

Example 2.8 *Suppose that $\{a_n\}_{n=0}^{\infty}$ is a period-2 sequence. Determine the pattern of the unique period-2 cycle of Eq. (2.21).*

Solution: *Set $x_2 = x_0$ and we acquire:*

$$\begin{aligned}
&x_0, \\
&x_1 = -a_0 x_0 + 1, \\
&x_2 = -a_1 x_1 + 1 = a_0 a_1 x_0 - a_1 + 1 = x_0.
\end{aligned} \tag{2.22}$$

Provided that $a_0 a_1 \neq 1$, via (2.22) we obtain the cognate initial condition:

$$x_0 = \frac{1 - a_1}{1 - a_0 a_1}, \tag{2.23}$$

and the associated **unique period-2 cycle***:*

$$\frac{1 - a_1}{1 - a_0 a_1}, \ \frac{1 - a_0}{1 - a_0 a_1}, \ \ldots \tag{2.24}$$

Observe that the coefficients of the sequence $\{a_n\}_{n=0}^{\infty}$ shift forward by an index of 1 in the numerator only. The subsequent sketch (Figure 2.18) renders (2.24) as a positive **decreasing period-2 cycle** *when $a_0 = 2$, $a_1 = 3$ and $x_0 = 0.4$:*

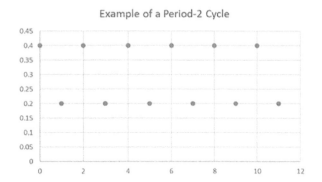

FIGURE 2.18: A Positive Decreasing Period-2 Cycle.

Example 2.9 *Suppose that $\{a_n\}_{n=0}^{\infty}$ is period-3 sequence. Determine the pattern of the unique period-3 cycle of Eq. (2.21).*

Solution: *Set $x_3 = x_0$ and we procure:*

$$
\begin{aligned}
&x_0, \\
&x_1 = -a_0 x_0 + 1, \\
&x_2 = -a_1 x_1 + 1 = a_0 a_1 x_0 - a_1 + 1, \\
&x_3 = -a_1 x_2 + 1 = -a_0 a_1 a_2 x_0 + a_1 a_2 - a_2 + 1 = x_0.
\end{aligned}
\tag{2.25}
$$

Provided that $a_0 a_1 a_2 \neq -1$, via (2.25) we acquire the corresponding initial condition:

$$
x_0 = \frac{a_1 a_2 - a_2 + 1}{1 + a_0 a_1 a_2}
\tag{2.26}
$$

and the related **unique period-3 cycle**:

$$
\frac{a_1 a_2 - a_2 + 1}{1 + a_0 a_1 a_2}, \ \frac{a_2 a_0 - a_0 + 1}{1 + a_0 a_1 a_2}, \ \frac{a_0 a_1 - a_1 + 1}{1 + a_0 a_1 a_2}, \dots.
\tag{2.27}
$$

The succeeding diagram (Figure 2.19) traces (2.27) as an **ascending triangular-shaped** *period-3 cycle with three positive terms when $a_0 = 1$, $a_1 = 1$, $a_2 = 4$ and $x_0 = 0.2$:*

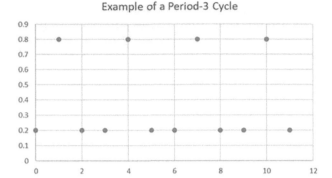

FIGURE 2.19: Ascending Triangular-Shaped Period-3 Cycle.

We can see quite contrasting differences between (2.23) and (2.26) and analogous disparities between (2.24) and (2.27). From Example (2.8), outlining the pattern of the initial condition and the pattern of the unique period-2k cycle ($k \in \mathbb{N}$) when $\{a_n\}_{n=0}^{\infty}$ is period-2k sequence will be left as an end-of-chapter exercise. Equivalently, from Example (2.9), rendering the pattern of the initial condition and the pattern of the unique period-(2k+1) cycle ($k \in \mathbb{N}$) when $\{a_n\}_{n=0}^{\infty}$ is period-(2k+1) sequence will be left as an end-of-chapter exercise.

We will adjourn this chapter with the periodic features of the corresponding difference equation in the additive and in the multiplicative form:

$$x_{n+1} = a_n x_n + b_n, \quad n = 0, 1, \ldots, \tag{2.28}$$

where $\{a_n\}_{n=0}^{\infty}$ and $\{b_n\}_{n=0}^{\infty}$ are periodic sequences with either the same period or different periods. Determining the unique periodic traits of Eq. (2.28) will be left as end-of-chapter exercises.

Piece-wise difference equations in Chapter 4 will be assembled as fragments of first-order linear difference equations.

2.3   Chapter 2 Exercises

In problems 1–4, determine the periodic cycle by solving the given **Initial Value Problem**:

1. $\begin{cases} x_{n+1} = -x_n, & n = 0, 1, \ldots. \\ \quad x_0 = -4. \end{cases}$

2. $\begin{cases} x_{n+1} = -x_n, & n = 0, 1, \ldots. \\ \quad x_0 = 2. \end{cases}$

3. $\begin{cases} x_{n+1} = -x_n + 4, & n = 0, 1, \dots. \\ x_0 = -2. \end{cases}$

4. $\begin{cases} x_{n+1} = -x_n + 4, & n = 0, 1, \dots. \\ x_0 = 6. \end{cases}$

5. $\begin{cases} x_{n+1} = (-1)^{n+1} x_n, & n = 0, 1, \dots. \\ x_0 = -3. \end{cases}$

6. $\begin{cases} x_{n+1} = (-1)^n x_n, & n = 0, 1, \dots. \\ x_0 = 3. \end{cases}$

7. $\begin{cases} x_{n+1} = (-1)^n x_n + 1, & n = 0, 1, \dots. \\ x_0 = -2. \end{cases}$

8. $\begin{cases} x_{n+1} = (-1)^n x_n - 1, & n = 0, 1, \dots. \\ x_0 = 6. \end{cases}$

In problems 9–12, using Examples (2.1–2.2) and (2.9), determine the periodic pattern of the following $\Delta.E.$:

$$x_{n+1} = a_n x_n, \quad n = 0, 1, \dots .$$

9. $\{a_n\}_{n=0}^{\infty}$ is a period-2 sequence and $a_0 a_1 = -1$.

10. $\{a_n\}_{n=0}^{\infty}$ is a period-3 sequence and $a_0 a_1 a_2 = -1$.

11. $\{a_n\}_{n=0}^{\infty}$ is a period-4 sequence and $a_0 a_1 a_2 a_3 = -1$.

12. $\{a_n\}_{n=0}^{\infty}$ is a period-k sequence ($k \geq 2$) and $\prod_{i=0}^{k-1} a_i = -1$.

In problems 13–16, determine the periodic traits and patterns of the following $\Delta.E.$:

$$x_{n+1} = a_n x_n + 1, \quad n = 0, 1, \dots,$$

13. when $\{a_n\}_{n=0}^{\infty}$ is a period-2 sequence (period-2 cycle).

14. when $\{a_n\}_{n=0}^{\infty}$ is a period-3 sequence (period-3 cycle).

15. when $\{a_n\}_{n=0}^{\infty}$ is a period-4 sequence (period-4 cycle).

16. when $\{a_n\}_{n=0}^{\infty}$ is a period-k sequence, ($k \geq 2$) (period-k cycle).

In problems 17–20, determine the periodic traits and patterns of the following $\Delta.E.$:

$$x_{n+1} = -a_n x_n + 1, \quad n = 0, 1, \dots,$$

17. when $\{a_n\}_{n=0}^{\infty}$ is a period-4 sequence (period-4 cycle).

18. when $\{a_n\}_{n=0}^{\infty}$ is a period-2k sequence, ($k \geq 1$) (period-2k cycle).

19. when $\{a_n\}_{n=0}^{\infty}$ is a period-5 sequence (period-5 cycle).

20. when $\{a_n\}_{n=0}^{\infty}$ is a period-$(2k+1)$ sequence, $(k \geq 1)$ (period-$(2k+1)$ cycle).

In problems $21 - 24$, determine the periodic traits and patterns of the following $\Delta.E.$:

$$x_{n+1} = a_n x_n + (-1)^n, \quad n = 0, 1, \dots,$$

21. when $\{a_n\}_{n=0}^{\infty}$ is a period-2 sequence (period-2 cycle).

22. when $\{a_n\}_{n=0}^{\infty}$ is a period-3 sequence (period-3 cycle).

23. when $\{a_n\}_{n=0}^{\infty}$ is a period-4 sequence (period-4 cycle).

24. when $\{a_n\}_{n=0}^{\infty}$ is a period-2k sequence, $(k \geq 1)$ (period-k cycle).

In problems 25–34, determine the pattern of the unique periodic cycle of:

$$x_{n+1} = a_n x_n + b_n, \quad n = 0, 1, \dots,$$

25. when $\{a_n\}_{n=0}^{\infty}$ and $\{b_n\}_{n=0}^{\infty}$ are period-2 sequences (period-2 cycle).

26. when $\{a_n\}_{n=0}^{\infty}$ and $\{b_n\}_{n=0}^{\infty}$ are period-3 sequences (period-3 cycle).

27. when $\{a_n\}_{n=0}^{\infty}$ and $\{b_n\}_{n=0}^{\infty}$ are period-4 sequences (period-4 cycle).

28. when $\{a_n\}_{n=0}^{\infty}$ and $\{b_n\}_{n=0}^{\infty}$ are period-k sequences, $(k \geq 2)$ (period-k cycle).

29. when $\{a_n\}_{n=0}^{\infty}$ is a period-2 sequence and $\{b_n\}_{n=0}^{\infty}$ is a period-4 sequence (period-4 cycle).

30. when $\{a_n\}_{n=0}^{\infty}$ is a period-4 sequence and $\{b_n\}_{n=0}^{\infty}$ is a period-2 sequence (period-4 cycle).

31. when $\{a_n\}_{n=0}^{\infty}$ is a period-3 sequence and $\{b_n\}_{n=0}^{\infty}$ is a period-6 sequence (period-6 cycle).

32. when $\{a_n\}_{n=0}^{\infty}$ is a period-6 sequence and $\{b_n\}_{n=0}^{\infty}$ is a period-3 sequence (period-6 cycle).

33. when $\{a_n\}_{n=0}^{\infty}$ is a period-2 sequence and $\{b_n\}_{n=0}^{\infty}$ is a period-3 sequence (period-6 cycle).

34. when $\{a_n\}_{n=0}^{\infty}$ is a period-3 sequence and $\{b_n\}_{n=0}^{\infty}$ is a period-2 sequence (period-6 cycle).

Chapter 3

Riccati Difference Equations

This chapter's goal is to get acquainted with numerous periodic traits of the mth-order Autonomous Riccati Difference Equation in the form:

$$x_{n+m} = \frac{A}{x_n}, \quad n = 0, 1, \ldots, \tag{3.1}$$

where $A > 0$ and $m \in \mathbb{N}$. In addition, we will examine the periodic traits of the mth-order Non-Autonomous Riccati $\Delta.$E. in the form:

$$x_{n+m} = \frac{A_n}{x_n}, \quad n = 0, 1, \ldots, \tag{3.2}$$

where $\{A_n\}_{n=0}^{\infty}$ is a period-k sequence ($k \geq 2$). The upcoming sections will examine the periodic features of Eqs. (3.1) and (3.2) when $m = 1$ and $m = 2$.

3.1 First-Order Riccati $\Delta.E.$

We will break off this section with the periodic character of the following first-order Riccati $\Delta.$E. (when $m = 1$):

$$x_{n+1} = \frac{A}{x_n}, \quad n = 0, 1, \ldots, \tag{3.3}$$

where $A > 0$ and the corresponding first-order Non-Autonomous Riccati $\Delta.$E. (when $m = 1$):

$$x_{n+1} = \frac{A_n}{x_n}, \quad n = 0, 1, \ldots, \tag{3.4}$$

where $\{A_n\}_{n=0}^{\infty}$ is a period-k sequence ($k \geq 2$). We will proceed with the periodic traits of Eq. (3.3). Observe that every solution of Eq. (3.3) is periodic with the corresponding period-2 pattern:

$$x_0,$$

$$x_1 = \frac{A}{x_0},$$

$$x_2 = \frac{A}{x_1} = \frac{A}{\frac{A}{x_0}} = x_0,$$

$$x_3 = \frac{A}{x_2} = \frac{A}{x_0} = x_1, \qquad (3.5)$$

$$\vdots$$

The consequent sketch (Figure 3.1) renders an increasing period-2 cycle of Eq. (3.3) when $x_0 = 0.8$ and $A = 2$:

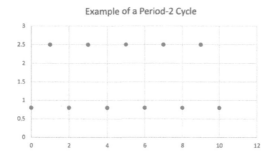

FIGURE 3.1: Riccati Period-2 Cycle.

Now we will transition to analyzing the periodic traits of Eq. (3.4). It is our goal to depict the assortment of combinations of the periodic patterns depending on the period of the period-k sequence $\{A_n\}_{n=0}^{\infty}$ and on the relationship of the sequence's terms. We will evince instances where the periodic cycles of Eq. (3.4) are unique and where every solution of Eq. (3.4) is periodic. We will discover these contrasts when $\{A_n\}_{n=0}^{\infty}$ is an even-ordered sequence in comparison to when $\{A_n\}_{n=0}^{\infty}$ is an odd-ordered sequence.

Example 3.1 *Suppose that $\{A_n\}_{n=0}^{\infty}$ is a period-2 sequence. Show that Eq. (3.4) has no period-2 cycles but has period-4 cycles and explain why.*

Solution: *We will show the existence of period-4 cycles and no existence of period-2 cycles. By iterations we procure:*

$$x_0,$$

$$x_1 = \frac{A_0}{x_0},$$

$$x_2 = \frac{A_1}{x_1} = \frac{A_1}{\left[\frac{A_0}{x_0}\right]} = \frac{A_1 x_0}{A_0},$$

$$x_3 = \frac{A_0}{x_2} = \frac{A_0}{\left[\frac{A_1 x_0}{A_0}\right]} = \frac{A_0^2}{A_1 x_0}, \qquad (3.6)$$

$$x_4 = \frac{A_1}{x_3} = \frac{A_1}{\left[\frac{A_0^2}{A_1 x_0}\right]} = \frac{A_1^2 x_0}{A_0^2}.$$

First of all, via (3.6) $x_2 = x_0$ if and only if $A_0 = A_1$. This contradicts the assumption that $\{A_n\}_{n=0}^{\infty}$ is a period-2 sequence where $A_0 \neq A_1$. Thus we conclude that Eq. (3.4) has no period-2 cycles. Second of all, via (3.6) $x_4 = x_0$ if and only if $A_0 = -A_1$ and this does not contradict that $\{A_n\}_{n=0}^{\infty}$ is a period-2 sequence. Hence Eq. (3.4) has period-4 cycles if and only if $A_0 = -A_1$. If fact, suppose that $A_1 = -A_0$, then via (3.6) we procure the following **alternating** *period-4 pattern:*

$$x_0, \quad \frac{A_0}{x_0}, \quad -x_0, \quad -\frac{A_0}{x_0}, \quad \dots \quad (3.7)$$

Now we will portray assorted shapes of the alternating period-4 cycle resembled in (3.7). The consequent sketch (Figure 3.2) renders (3.7) as a **descending step-shaped** *alternating period-4 cycle when $x_0 = 2$, $A_0 = 4$ and $A_1 = -4$:*

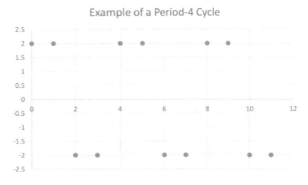

FIGURE 3.2: Riccati Period-4 Cycle.

The next diagram (Figure 3.3) traces (3.7) as an **ascending trapezoidal-shaped** *alternating period-4 cycle when $x_0 = -3$, $A_0 = -9$ and $A_1 = 9$:*

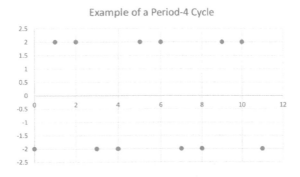

FIGURE 3.3: Riccati Period-4 Cycle.

Example 3.2 *Suppose that $\{A_n\}_{n=0}^{\infty}$ is a period-3 sequence. Determine the existence, uniqueness and the pattern of the period-3 cycles of Eq. (3.4).*

Solution: *Assume that $x_3 = x_0$ and we acquire:*

$$x_0,$$

$$x_1 = \frac{A_0}{x_0},$$

$$x_2 = \frac{A_1}{x_1} = \frac{A_1}{\left[\frac{A_0}{x_0}\right]} = \frac{A_1 x_0}{A_0},$$

$$x_3 = \frac{A_2}{x_2} = \frac{A_2}{\left[\frac{A_1 x_0}{A_0}\right]} = \frac{A_2 A_0}{A_1 x_0} = x_0.$$

Notice that either

$$x_0 = \sqrt{\frac{A_0 A_2}{A_1}} \quad \text{or} \quad x_0 = -\sqrt{\frac{A_0 A_2}{A_1}}. \tag{3.8}$$

Now we will verify the existence of the unique positive period-3 cycle:

$$x_0 = \sqrt{\frac{A_0 A_2}{A_1}},$$

$$x_1 = \frac{A_0}{x_0} = \frac{A_0}{\sqrt{\frac{A_2 A_0}{A_1}}} = \sqrt{\frac{A_1 A_0}{A_2}},$$

$$x_2 = \frac{A_1}{x_1} = \frac{A_1}{\sqrt{\frac{A_0 A_1}{A_2}}} = \sqrt{\frac{A_2 A_1}{A_0}},$$

$$x_3 = \frac{A_2}{x_2} = \frac{A_2}{\sqrt{\frac{A_1 A_2}{A_0}}} = \sqrt{\frac{A_0 A_2}{A_1}} = x_0.$$

Hence we see the unique positive period-3 cycle:

$$\sqrt{\frac{A_0 A_2}{A_1}}, \ \sqrt{\frac{A_1 A_0}{A_2}}, \ \sqrt{\frac{A_2 A_1}{A_0}}, \ \ldots \tag{3.9}$$

Analogous to (3.9), we acquire the corresponding unique negative period-3 cycle:

$$-\sqrt{\frac{A_0 A_2}{A_1}}, \ -\sqrt{\frac{A_1 A_0}{A_2}}, \ -\sqrt{\frac{A_2 A_1}{A_0}}, \ \ldots$$

*The consequent diagram (Figure 3.4) depicts a **descending step-shaped** period-3 cycle of Eq. (3.4) when $x_0 = 2$, $A_0 = 2$, $A_1 = 1$ and $A_2 = 2$:*

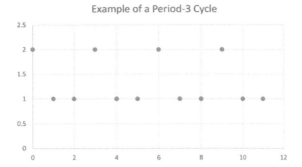

FIGURE 3.4: Riccati Descending Step-Shaped Period-3 Cycle.

In all other cases, Eq. (3.4) is periodic with period-6. In fact, the corresponding sketch (Figure 3.5) renders a period-6 cycle when $x_0 = 1$, $A_0 = 2$, $A_1 = 1$ and $A_2 = 1$:

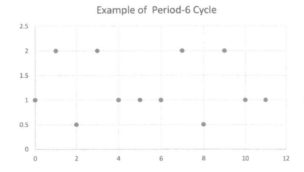

FIGURE 3.5: Riccati Period-6 Cycle.

Now suppose that $\{A_n\}_{n=0}^{\infty}$ is a period-5 sequence. From Examples (3.2) and (3.8), we acquire the following initial condition:

$$x_0 = \sqrt{\frac{A_0 A_2 A_4}{A_1 A_3}}, \tag{3.10}$$

which produces the unique positive period-5 cycle of Eq. (3.4). The succeeding figure (Figure 3.6) renders the unique positive period-5 cycle of Eq. (3.4) when $x_0 = 4$, $A_0 = 2$, $A_1 = 1$, $A_2 = 2$, $A_3 = 1$ and $A_4 = 4$:

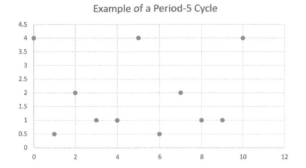

FIGURE 3.6: Riccati Period-5 Cycle.

Similar to Example (3.2), in all other cases every solution of Eq. (3.4) is periodic with period-10. From Example (3.2), we can conclude that Eq. (3.4) has unique periodic cycles when $\{A_n\}_{n=0}^{\infty}$ is an odd-ordered period.

Theorem 3.1 *Suppose that $\{A_n\}_{n=0}^{\infty}$ is a periodic sequence with period-$(2k+1)$, $(k \in \mathbb{N})$. Then Eq. (3.4) has unique period-$(2k+1)$ cycles where either:*

$$x_0 = \sqrt{\frac{\prod_{i=1}^{k+1} A_{2i-2}}{\prod_{i=1}^{k} A_{2i-1}}} \quad \text{or} \quad x_0 = -\sqrt{\frac{\prod_{i=1}^{k+1} A_{2i-2}}{\prod_{i=1}^{k} A_{2i-1}}}. \tag{3.11}$$

In all other cases, every solution of Eq. (3.4) is periodic period-$2(2k+1)$.

The pattern of (3.11) emerges on the base of (3.8) and (3.10). The proof of (3.11) is obtained by induction and will be left as an end-of-chapter exercise. Now we will direct our focus on when $\{A_n\}_{n=0}^{\infty}$ is an even-ordered periodic sequence.

Example 3.3 *Suppose that $\{A_n\}_{n=0}^{\infty}$ is a period-4 sequence. Determine the existence, uniqueness and the pattern of the period-4 cycles of Eq. (3.4).*

Solution: *Analogous to Example (3.2), let $x_4 = x_0$ and we acquire:*

$$x_0,$$

$$x_1 = \frac{A_0}{x_0},$$

$$x_2 = \frac{A_1}{x_1} = \frac{A_1}{\left[\frac{A_0}{x_0}\right]} = \frac{A_1 x_0}{A_0},$$

$$x_3 = \frac{A_2}{x_2} = \frac{A_2}{\left[\frac{A_1 x_0}{A_0}\right]} = \frac{A_2 A_0}{A_1 x_0}, \tag{3.12}$$

$$x_4 = \frac{A_3}{x_3} = \frac{A_3}{\left[\frac{A_2 a_0}{A_1 x_0}\right]} = \frac{A_3 A_1 x_0}{A_2 A_0} = x_0.$$

We see that $x_4 = x_0$ if and only if

$$A_3 A_1 = A_2 A_0. \tag{3.13}$$

Therefore, every solution of Eq. (3.4) is periodic with the following period-4 pattern:

$$x_0, \ \frac{A_0}{x_0}, \ \frac{A_1 x_0}{A_0}, \ \frac{A_2 A_0}{A_1 x_0}, \ \dots$$

Via (3.12) we conclude that Eq. (3.4) has no period-2 cycles as:

(1) $x_2 = \frac{A_1 x_0}{A_0} = x_0$ if and only if $A_0 = A_1$,

(2) $x_3 = \frac{A_2 A_0}{A_1 x_0} = \frac{A_0}{x_0}$ if and only if $A_1 = A_2$,

(3) $x_4 = \frac{A_3 A_1 x_0}{A_2 A_0} = \frac{A_1 x_0}{A_0}$ if and only if $A_2 = A_3$.

Thus Eq. (3.4) has a period-2 cycle if and only if $A_0 = A_1 = A_2 = A_3$ which is a contraction as we assumed that $\{A_n\}_{n=0}^{\infty}$ is a period-4 sequence.

*The upcoming sketch (Figure 3.7) renders a **descending step-shaped** period-4 cycle of Eq. (3.4) when $x_0 = 2$, $A_0 = 4$, $A_1 = 2$, $A_2 = 1$ and $A_3 = 2$:*

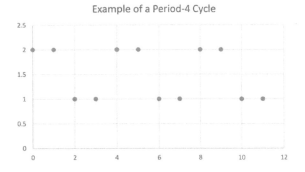

FIGURE 3.7: Riccati Descending Step-Shaped Period-4 Cycle.

Now assume that $\{A_n\}_{n=0}^{\infty}$ is a periodic sequence with period-2k, $(k \geq 2)$. The upcoming theorem extends the results from Example (3.3).

Theorem 3.2 *Suppose that $\{A_n\}_{n=0}^{\infty}$ is a periodic sequence with period-2k, $(k \geq 2)$. Then every solution of Eq. (3.4) is periodic with period-2k if and only if:*

$$\prod_{i=1}^{k} a_{2i-2} = \prod_{i=1}^{k} a_{2i-1}. \tag{3.14}$$

Notice that (3.14) develops from (3.13). Also the left-hand side of (3.14) is the product of all the even-indexed terms of $\{A_n\}_{n=0}^{\infty}$, while the right-hand side of (3.14) is the product of all the odd-indexed terms of $\{A_n\}_{n=0}^{\infty}$. The proof of (3.14) is done by induction and will be left as an end-of-chapter exercise. The next section will transition to the second-order Autonomous and Non-Autonomous Riccati Difference Equations when $m = 2$.

3.2   Second-Order Riccati $\Delta.E.$

We will focus on the periodic traits of the second-order Riccati Δ.E. (when $m = 2$):

$$x_{n+2} = \frac{A}{x_n}, \quad n = 0, 1, \ldots, \tag{3.15}$$

where $A > 0$ and the second-order Non-Autonomous Riccati Δ.E.:

$$x_{n+2} = \frac{A_n}{x_n}, \quad n = 0, 1, \ldots, \tag{3.16}$$

where $\{A_n\}_{n=0}^{\infty}$ is a period-k sequence $(k \geq 2)$. We will commence with the periodic features of Eq. (3.15) in comparison to Eq. (3.3). Analogous to (3.5), every solution of Eq. (3.15) is periodic with the related period-4 pattern:

$$x_0, \ x_1, \ \frac{A}{x_0}, \ \frac{A}{x_1}, \ \ldots \tag{3.17}$$

The period-4 pattern of (3.17) mimics the period-2 pattern of (3.5). The upcoming diagram (Figure 3.8) depicts an **ascending trapezoidal-shaped period-4 cycle** of Eq. (3.15) when $x_0 = 2$, $x_1 = 4$ and $A = 7$:

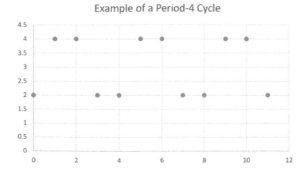

FIGURE 3.8: Trapezoidal-Shaped Riccati Period-4 Cycle.

The patterns and shapes of the period-4 cycles of Eq. (3.17) vary depending on the values of x_0, x_1 and A in contrast to either increasing or decreasing period-2 patterns of Eq. (3.3).

Now we will advance with investigating the periodic traits of Eq. (3.16). Our aim is to describe the variety of combinations of the periodic patterns depending on the period of the sequence $\{A_n\}_{n=0}^{\infty}$ and on the arrangement of the sequence's terms. Analogous to the periodicity properties of Eq. (3.4), we will study various occurrences when the periodic cycles of Eq. (3.16) are unique and when every solution of Eq. (3.16) is periodic. We will discover these disparities when $\{A_n\}_{n=0}^{\infty}$ is an even-ordered sequence in comparison to when $\{A_n\}_{n=0}^{\infty}$ is an odd-ordered sequence.

Example 3.4 *Suppose that $\{A_n\}_{n=0}^{\infty}$ is a period-2 sequence. Determine the periodic traits of Eq. (3.16).*

Solution: *We will show the existence of unique period-2 cycles and the existence of period-4 cycles. First suppose that $x_2 = x_0$ and $x_3 = x_1$ and we obtain:*

$$x_0,$$
$$x_1,$$
$$x_2 = \frac{A_0}{x_0} = x_0, \tag{3.18}$$
$$x_3 = \frac{A_1}{x_1} = x_1.$$

Note that via (3.18) we procure:

$$x_0 = \sqrt{A_0} \quad \text{and} \quad x_1 = \sqrt{A_1}, \tag{3.19}$$

and the corresponding unique period-2 cycle:

$$\sqrt{A_0}, \ \sqrt{A_1}, \ \sqrt{A_0}, \ \sqrt{A_1}, \ \dots \tag{3.20}$$

*The consequent graph (Figure 3.9) evokes (3.20) as a **descending** period-2 cycle of Eq. (3.16) when $x_0 = 4$, $x_1 = 1$, $A_0 = 16$ and $A_1 = 1$:*

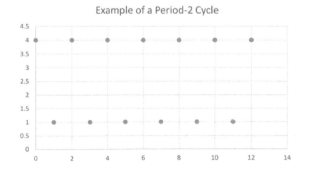

FIGURE 3.9: Second-Order Riccati Descending Period-2 Cycle.

In all other cases, via (3.18) every solution of Eq. (3.16) is periodic with the corresponding period-4 cycle:

$$x_0, \; x_1, \; \frac{A_0}{x_0}, \; \frac{A_1}{x_1}, \; \dots \quad (3.21)$$

Notice that (3.21) can be resembled in numerous configurations depending on the chosen values of x_0, x_1, A_0 and A_1 such as ascending, descending, step-shaped, trapezoidal-shaped and additional patterns.

The succeeding two graphs will render different periodic shapes that trace the pattern of (3.21). The upcoming sketch (Figure 3.10) traces a **descending step-shaped** *period-4 cycle of Eq. (3.16) when $x_0 = 4$, $x_1 = 3$, $A_0 = 8$ and $A_1 = 6$:*

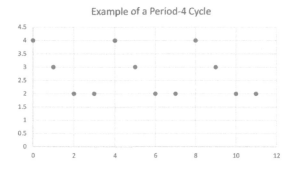

FIGURE 3.10: Descending Step-Shaped Second-Order Riccati Period-4 Cycle.

In contrast to Figure 3.10, by altering the values of x_0, x_1, A_0 and A_1, the diagram below (Figure 3.11) depicts an **ascending step-shaped** *period-4 cycle of Eq. (3.16) when $x_0 = 1$, $x_1 = 1$, $A_0 = 2$ and $A_1 = 4$:*

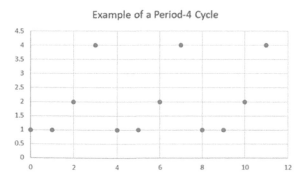

FIGURE 3.11: Ascending Step-Shaped Second-Order Riccati Period-4 Cycle.

Example 3.5 *Suppose that $\{A_n\}_{n=0}^{\infty}$ is a period-3 sequence. Determine the periodic traits of Eq. (3.16).*

Solution: *We will show the existence of unique period-3 cycles and the existence of period-12 cycles. By iterations we procure:*

$$x_0,$$

$$x_1,$$

$$x_2 = \frac{A_0}{x_0},$$

$$x_3 = \frac{A_1}{x_1},$$

$$x_4 = \frac{A_2}{x_2} = \frac{A_2}{\left[\frac{A_0}{x_0}\right]} = \frac{A_2 x_0}{A_0},$$

$$x_5 = \frac{A_0}{x_3} = \frac{A_0}{\left[\frac{A_1}{x_1}\right]} = \frac{A_0 x_1}{A_1},$$

$$x_6 = \frac{A_1}{x_4} = \frac{A_1}{\left[\frac{A_2 x_0}{A_0}\right]} = \frac{A_0 A_1}{A_2 x_0},$$

$$x_7 = \frac{A_2}{x_5} = \frac{A_2}{\left[\frac{A_0 x_1}{A_1}\right]} = \frac{A_2 A_1}{A_0 x_1},$$

$$x_8 = \frac{A_0}{x_6} = \frac{A_0}{\left[\frac{A_0 A_1}{A_2 x_0}\right]} = \frac{A_2 x_0}{A_1},$$

$$x_9 = \frac{A_1}{x_7} = \frac{A_1}{\left[\frac{A_2 A_1}{A_0 x_1}\right]} = \frac{A_0 x_1}{A_2}, \tag{3.22}$$

$$x_{10} = \frac{A_2}{x_8} = \frac{A_2}{\left[\frac{A_2 x_0}{A_1}\right]} = \frac{A_1}{x_0},$$

$$x_{11} = \frac{A_0}{x_9} = \frac{A_0}{\left[\frac{A_0 x_1}{A_2}\right]} = \frac{A_2}{x_1},$$

$$x_{12} = \frac{A_1}{x_{10}} = \frac{A_1}{\left[\frac{A_1}{x_0}\right]} = x_0,$$

$$x_{13} = \frac{A_2}{x_{11}} = \frac{A_2}{\left[\frac{A_2}{x_1}\right]} = x_1.$$

First of all, to obtain period-3 cycles from (3.22) we set $x_6 = x_0$ and $x_7 = x_1$ and we acquire:

$$x_0 = \sqrt{\frac{A_0 A_1}{A_2}} \quad \text{and} \quad x_1 = \sqrt{\frac{A_1 A_2}{A_0}}, \tag{3.23}$$

and the corresponding unique period-3 pattern:

$$\sqrt{\frac{A_0 A_1}{A_2}}, \ \sqrt{\frac{A_1 A_2}{A_0}}, \ \sqrt{\frac{A_2 A_0}{A_1}}, \ \ldots \tag{3.24}$$

Depending on the values of A_0, A_1 and A_2, we can find assorted structures of period-3 cycles that render (3.24). The graph below (Figure 3.12) evokes a **descending triangular-shaped** *period-3 cycle of Eq. (3.16) when $x_0 = 2$, $x_1 = 0.5$, $A_0 = 4$, $A_1 = 1$ and $A_2 = 1$:*

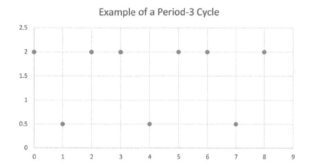

FIGURE 3.12: Triangular-Shaped Second-Order Riccati Period-3 Cycle.

On the contrary, we can detect an **ascending triangular-shaped** *period-3 cycle of Eq. (3.16) when $x_0 = 1$, $x_1 = 4$, $A_0 = 1$, $A_1 = 4$ and $A_2 = 4$ (Figure 3.13):*

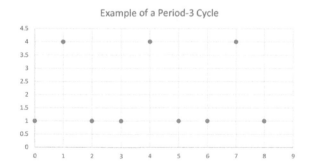

FIGURE 3.13: Triangular-Shaped Second-Order Riccati Period-3 Cycle.

For instance, the next sketch (Figure 3.14) traces a **descending step-shaped** *period-3 cycle of Eq. (3.16) when $x_0 = 4$, $x_1 = 1$, $A_0 = 4$, $A_1 = 4$ and $A_2 = 1$:*

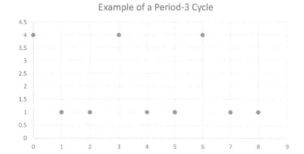

FIGURE 3.14: Descending Step-Shaped Second-Order Riccati Period-3 Cycle.

On the contrast with Figure 3.14, the upcoming diagram (Figure 3.15) depicts an **ascending step-shaped** *period-3 cycle of Eq. (3.16) when $x_0 = 1$, $x_1 = 1$, $A_0 = 6$, $A_1 = 1$ and $A_2 = 6$:*

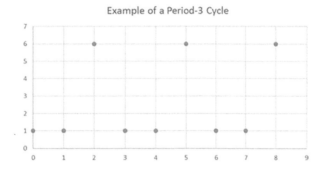

FIGURE 3.15: Ascending Step-Shaped Second-Order Riccati Period-3 Cycle.

Second, to obtain period-4 cycles from (3.22), we set $x_4 = x_0$ and $x_5 = x_1$ and acquire:

(1) $x_4 = \frac{A_2 x_0}{A_0} = x_0$ if and only if $A_0 = A_2$,

(2) $x_5 = \frac{A_0 x_1}{A_1} = x_1$ if and only if $A_0 = A_1$.

Hence Eq. (3.16) has a period-4 cycle if and only if $A_0 = A_1 = A_2$ which is a contraction as we assumed that $\{A_n\}_{n=0}^{\infty}$ is a period-3 sequence. Therefore, via (3.22) we see that in all other cases every solution of Eq. (3.16) is periodic with period-12. The next sketch (Figure 3.16) describes a period-12 cycle of Eq. (3.16) when $x_0 = 1$, $x_1 = 2$, $A_0 = 2$, $A_1 = 1$ and $A_2 = 1$:

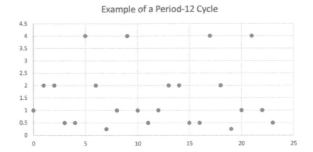

FIGURE 3.16: Second-Order Riccati Period-12 Cycle.

Now will assume that $\{A_n\}_{n=0}^{\infty}$ is a period-4 sequence and show that Eq. (3.16) has no periodic cycles. In fact, this will mimic Example (3.1).

Example 3.6 *Suppose that $\{A_n\}_{n=0}^{\infty}$ is a period-4 sequence. Show that Eq. (3.16) has no period-4 cycles but has period-8 cycles and explain why.*

Solution: *We will show the existence of period-8 cycles and no existence of period-4 cycles. By iterations we procure:*

$$x_0,$$

$$x_1,$$

$$x_2 = \frac{A_0}{x_0},$$

$$x_3 = \frac{A_1}{x_1},$$

$$x_4 = \frac{A_2}{x_2} = \frac{A_2}{\left[\frac{A_0}{x_0}\right]} = \frac{A_2 x_0}{A_0},$$

$$x_5 = \frac{A_3}{x_3} = \frac{A_3}{\left[\frac{A_1}{x_1}\right]} = \frac{A_3 x_1}{A_1},$$

$$x_6 = \frac{A_0}{x_4} = \frac{A_0}{\left[\frac{A_2 x_0}{A_0}\right]} = \frac{A_0^2}{A_2 x_0}, \qquad (3.25)$$

$$x_7 = \frac{A_1}{x_5} = \frac{A_1}{\left[\frac{A_3 x_1}{A_1}\right]} = \frac{A_1^2}{A_3 x_1},$$

$$x_8 = \frac{A_2}{x_6} = \frac{A_2}{\left[\frac{A_0^2}{A_2 x_0}\right]} = \frac{A_2^2 x_0}{A_0^2},$$

$$x_9 = \frac{A_3}{x_7} = \frac{A_3}{\left[\frac{A_1^2}{A_3 x_1}\right]} = \frac{A_3^2 x_1}{A_1^2}.$$

Analogous to Example (3.1), via (3.25) we see that $x_4 = x_0$ and $x_5 = x_1$ if and only if $A_0 = A_2$ and $A_1 = A_3$. This contradicts the assumption that $\{A_n\}_{n=0}^{\infty}$ is a period-4 sequence. Hence we conclude that Eq. (3.16) has no period-4 cycles. Second of all, via (3.25) $x_8 = x_0$ and $x_9 = x_1$ if and only if $A_0 = -A_2$ and $A_1 = -A_3$. This does not contradict that $\{A_n\}_{n=0}^{\infty}$ is a period-4 sequence. Thus Eq. (3.4) has period-8 cycles if and only if $A_0 = -A_2$ and $A_1 = -A_3$. Notice that via (3.25), when $A_0 = -A_2$ and $A_1 = -A_3$, we obtain the corresponding **alternating** *period-8 pattern:*

$$x_0, \ x_1, \ \frac{A_0}{x_0}, \ \frac{A_1}{x_1}, \ -x_0, \ -x_1, \ -\frac{A_0}{x_0}, \ -\frac{A_1}{x_1}, \ \ldots \tag{3.26}$$

The diagram below (Figure 3.17) traces an alternating period-8 cycle of Eq. (3.16) when $x_0 = 2$, $x_1 = 1$, $A_0 = 4$, $A_1 = 1$, $A_2 = -4$ and $A_3 = -1$:

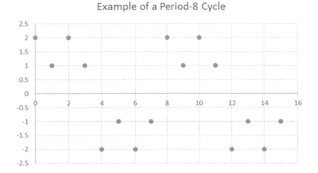

Example of a Period-8 Cycle

FIGURE 3.17: Second-Order Alternating Riccati Period-8 Cycle.

Now we will transition the study of periodic traits of Eq. (3.16) when $\{A_n\}_{n=0}^{\infty}$ is a period-5 sequence.

Example 3.7 *Suppose that $\{A_n\}_{n=0}^{\infty}$ is a period-5 sequence. Mimicking Example (3.5), we set $x_{10} = x_0$ and $x_{11} = x_1$ and we acquire:*

$$x_0 = \sqrt{\frac{A_3 A_4 A_0}{A_1 A_2}} \quad \text{and} \quad x_1 = \sqrt{\frac{A_4 A_0 A_1}{A_2 A_3}}, \tag{3.27}$$

which is similar to (3.23) and the corresponding unique period-5 pattern:

$$\sqrt{\frac{A_3 A_4 A_0}{A_1 A_2}}, \ \sqrt{\frac{A_4 A_0 A_1}{A_2 A_3}}, \ \sqrt{\frac{A_0 A_1 A_2}{A_3 A_4}}, \ \sqrt{\frac{A_1 A_2 A_3}{A_4 A_0}}, \ \sqrt{\frac{A_2 A_3 A_4}{A_0 A_1}}, \ \ldots, \tag{3.28}$$

whose arrangement resembles (3.24). Observe that the indices of the coefficients of the sequence $\{A_n\}_{n=0}^{\infty}$ shift forward by one from term to term. Alternatively, we can reformulate (3.28) as:

$$x_i = \sqrt{\frac{A_{(3+i)} A_{(4+i)} A_{(0+i)}}{A_{(1+i)} A_{(2+i)}}} \quad \text{for all} \quad i \in [0, 1, 2, 3, 4]. \tag{3.29}$$

In addition, analogous to Example (3.5) we can conclude that in all other cases, every solution of Eq. (3.16) is periodic with period-20.

The period-5 cycles rendering (3.28) can emerge in numerous shapes that may be similar to the periodic patterns we encountered in the previous examples. The upcoming sketch (Figure 3.18) describes a period-5 cycle of Eq. (3.16) when $x_0 = 2$, $x_1 = 8$, $A_0 = 4$, $A_1 = 4$, $A_2 = 1$, $A_3 = 1$ and $A_4 = 4$:

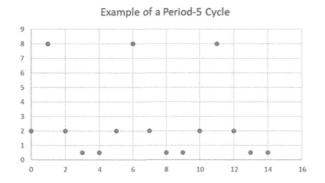

FIGURE 3.18: Second-Order Riccati Period-5 Cycle.

Next we will examine the periodic traits of Eq. (3.16) when $\{A_n\}_{n=0}^{\infty}$ is a period-6 sequence.

Example 3.8 *Suppose that $\{A_n\}_{n=0}^{\infty}$ is a period-6 sequence. Parallel to Example (3.4), Eq. (3.16) will render a unique period-6 cycle and period-12 cycles. To determine the unique period-6 cycle we set $x_6 = x_0$ and $x_7 = x_1$ and we obtain:*

$$x_0 = \sqrt{\frac{A_4 A_0}{A_2}} \quad \text{and} \quad x_1 = \sqrt{\frac{A_5 A_1}{A_3}}, \tag{3.30}$$

and the corresponding unique period-6 pattern:

$$x_i = \sqrt{\frac{A_{(4+i)} A_{(0+i)}}{A_{(2+i)}}} \quad \text{for all} \quad i \in [0, 1, 2, 3, 4, 5]. \tag{3.31}$$

Similar to the period-5 pattern in (3.29), the indices of the period-6 pattern in (3.31) also shift forward by 1 from term to term. The next diagram (Figure 3.19) portrays the unique period-6 cycle of Eq. (3.16) when $x_0 = 4$, $x_1 = 2$, $A_0 = 4$, $A_1 = 4$, $A_2 = 1$, $A_3 = 1$, $A_4 = 4$ and $A_5 = 1$:

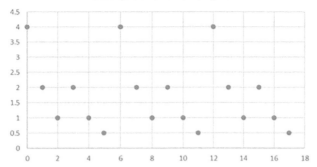

FIGURE 3.19: Second-Order Riccati Period-6 Cycle.

In all other cases, every solution of Eq. (3.16) is periodic with period-12.

Now assume that $\{A_n\}_{n=0}^{\infty}$ is a period-$(2k+1)$ sequence $(k \in \mathbb{N})$. The upcoming theorem will generalize the periodic traits of Eq. (3.16) from Examples (3.5) and (3.7).

Theorem 3.3 *Suppose that $\{A_n\}_{n=0}^{\infty}$ is a period-$(2k+1)$ sequence $(k \in \mathbb{N})$. Then exactly one of the following statements holds true:*

(1) Eq. (3.16) has a unique period-$(2k+1)$ cycle, or

(2) Every solution of Eq. (3.16) is periodic with period-$4(2k+1)$.

Proving theorem (3.3) will be left as an end-of-chapter exercise.

Next suppose that $\{A_n\}_{n=0}^{\infty}$ is a period-$(4k+2)$ sequence $(k \geq 0)$. The succeeding Theorem will extend the periodic features of Eq. (3.16) from Examples (3.4) and (3.8).

Theorem 3.4 *Suppose that $\{A_n\}_{n=0}^{\infty}$ is a period-$(4k+2)$ sequence $(k \geq 0)$. Then exactly one of the following statements holds true:*

(1) Eq. (3.16) has a unique period-$(2k+1)$ cycle, or

(2) Every solution of Eq. (3.16) is periodic with period-$(4k+2)$.

Proving Theorem (3.4) will be left as an end-of-chapter exercise. Determining the periodic attributes of Eq. (3.16) when $\{A_n\}_{n=0}^{\infty}$ is a period-$(4k)$ sequence $(k \geq 2)$ will be left as an end-of-chapter exercise.

Furthermore, examining the periodic features of Eqs. (3.1) and (3.2) when $m \geq 3$ and will be left as end-of-chapter exercises. In Chapter 5, we will assemble Max-Type Difference Equations as combinations of Autonomous and Non-Autonomous Riccati Difference Equations and encounter various periodic and eventually periodic cycles.

3.3   Chapter 3 Exercises

Consider the first-order **Non-Autonomous Riccati Δ.E.**:

$$x_{n+1} = \frac{A_n}{x_n}, \quad n = 0, 1, \ldots.$$

where $\{A_n\}_{n=0}^{\infty}$ is a periodic sequence. In problems 1–10:

1. Suppose that $\{A_n\}_{n=0}^{\infty}$ is a period-5 sequence. Determine the pattern of the periodic cycle(s).

2. Suppose that $\{A_n\}_{n=0}^{\infty}$ is a period-6 sequence. Determine the pattern of the periodic cycle(s).

3. Suppose that $\{A_n\}_{n=0}^{\infty}$ is a period-7 sequence. Determine the pattern of the periodic cycle(s).

4. Suppose that $\{A_n\}_{n=0}^{\infty}$ is a period-8 sequence. Determine the pattern of the periodic cycle(s).

5. Suppose that $\{A_n\}_{n=0}^{\infty}$ is a period-9 sequence. Determine the pattern of the periodic cycle(s).

6. Suppose that $\{A_n\}_{n=0}^{\infty}$ is a period-10 sequence. Determine the pattern of the periodic cycle(s).

7. Suppose that $\{A_n\}_{n=0}^{\infty}$ is a period-12 sequence. Determine the pattern of the periodic cycle(s).

8. Using problems 1, 3, and 5, suppose that $\{A_n\}_{n=0}^{\infty}$ is a period-$(2k+1)$ sequence, $(k \in \mathbb{N})$. Determine the pattern of the periodic cycle(s).

9. Using problems 2 and 6, suppose that $\{A_n\}_{n=0}^{\infty}$ is a period-$(4k+2)$ sequence, $(k \in \mathbb{N})$. Determine the pattern of the periodic cycle(s).

10. Using problems 4 and 7, suppose that $\{A_n\}_{n=0}^{\infty}$ is a period-4k sequence, $(k \in \mathbb{N})$. Determine the pattern of the periodic cycle(s).

Consider the second-order **Non-Autonomous Riccati Δ.E.**:

$$x_{n+2} = \frac{A_n}{x_n}, \quad n = 0, 1, \ldots.$$

where $\{A_n\}_{n=0}^{\infty}$ is a periodic sequence. In problems 11–20:

11. Suppose that $\{A_n\}_{n=0}^{\infty}$ is a period-5 sequence. Determine the pattern of the periodic cycle(s).

12. Suppose that $\{A_n\}_{n=0}^{\infty}$ is a period-6 sequence. Determine the pattern of the periodic cycle(s).

13. Suppose that $\{A_n\}_{n=0}^{\infty}$ is a period-7 sequence. Determine the pattern of the periodic cycle(s).

14. Suppose that $\{A_n\}_{n=0}^{\infty}$ is a period-8 sequence. Determine the pattern of the periodic cycle(s).

15. Suppose that $\{A_n\}_{n=0}^{\infty}$ is a period-9 sequence. Determine the pattern of the periodic cycle(s).

16. Suppose that $\{A_n\}_{n=0}^{\infty}$ is a period-10 sequence. Determine the pattern of the periodic cycle(s).

17. Suppose that $\{A_n\}_{n=0}^{\infty}$ is a period-12 sequence. Determine the pattern of the periodic cycle(s).

18. Using problems 11, 13, and 15, suppose that $\{A_n\}_{n=0}^{\infty}$ is a period-$(2k+1)$ sequence, $(k \in \mathbb{N})$. Determine the pattern of the periodic cycle(s).

19. Using problems 12 and 16, suppose that $\{A_n\}_{n=0}^{\infty}$ is a period-$(4k+2)$ sequence, $(k \in \mathbb{N})$. Determine the pattern of the periodic cycle(s).

20. Using problems 14 and 17, suppose that $\{A_n\}_{n=0}^{\infty}$ is a period-$4k$ sequence, $(k \in \mathbb{N})$. Determine the pattern of the periodic cycle(s).

Consider the third-order **Non-Autonomous Riccati** Δ.**E.**:

$$x_{n+3} = \frac{A_n}{x_n}, \quad n = 0, 1, \ldots.$$

where $\{A_n\}_{n=0}^{\infty}$ is a periodic sequence. In problems 21–26:

21. Suppose that $\{A_n\}_{n=0}^{\infty}$ is a period-2 sequence. Determine the pattern of the periodic cycle(s).

22. Suppose that $\{A_n\}_{n=0}^{\infty}$ is a period-3 sequence. Determine the pattern of the periodic cycle(s).

23. Suppose that $\{A_n\}_{n=0}^{\infty}$ is a period-4 sequence. Determine the pattern of the periodic cycle(s).

24. Suppose that $\{A_n\}_{n=0}^{\infty}$ is a period-5 sequence. Determine the pattern of the periodic cycle(s).

25. Suppose that $\{A_n\}_{n=0}^{\infty}$ is a period-6 sequence. Determine the pattern of the periodic cycle(s).

26. Suppose that $\{A_n\}_{n=0}^{\infty}$ is a period-7 sequence. Determine the pattern of the periodic cycle(s).

Chapter 4

Piece-wise Difference Equations

This chapter's aim is to investigate the existence of periodic solutions and eventually periodic solutions of first-order Piece-wise Difference Equations. Analogous to a Piece-wise function that is assembled from two or more functions such as:

$$y = \begin{cases} x & \text{if } x \leq 2 \\ 4 - x & \text{if } x > 4, \end{cases}$$

a **Piece-wise Difference Equation** is a recursive sequence that consists of two or more fragments. The following are examples of Piece-wise Difference Equations that we will thoroughly examine throughout this chapter:

(i) The Collatz Equation:

$$x_{n+1} = \begin{cases} \frac{x_n}{2} & \text{if } x_n \text{ is even,} \\ 3x_n + 1 & \text{if } x_n \text{ is odd,} \end{cases} \quad n = 0, 1, \ldots,$$

(ii) The 3X+1 Conjecture:

$$x_{n+1} = \begin{cases} \frac{x_n}{2} & \text{if } x_n \text{ is even,} \\ \frac{3x_n+1}{2} & \text{if } x_n \text{ is odd,} \end{cases} \quad n = 0, 1, \ldots,$$

(iii) Tent-Map:

$$x_{n+1} = \begin{cases} 2x_n & \text{if } x_n < \frac{1}{2}, \\ 2(1 - x_n) & \text{if } x_n \geq \frac{1}{2}. \end{cases} \quad n = 0, 1, \ldots.$$

(iv) Neuron Model:

$$x_{n+1} = \beta x_n - g(x_n), \quad n = 0, 1, 2, \ldots,$$

where

$$g(x) = \begin{cases} 1 & \text{if } x \geq 0, \\ -1 & \text{if } x < 0. \end{cases}$$

The Collatz Equation in (i) and the 3x+1 Conjecture in (ii) are special cases of the Collatz Conjectures.

63

4.1   The Collatz Conjectures

This section will examine the **Collatz Conjectures**, which are special cases of Piece-wise Difference Equations. We will commence with the following **Collatz Difference Equation**:

$$x_{n+1} = \begin{cases} \frac{x_n}{2} & \text{if } x_n \text{ is even,} \\ 3x_n + 1 & \text{if } x_n \text{ is odd,} \end{cases} \quad n = 0, 1, \ldots, \tag{4.1}$$

where $x_0 \in \mathbb{N}$. The initial condition $x_0 = 1$ produces the following unique period-3 pattern:

$$1, \ 4, \ 2, \ 1, \ 4, \ 2, \ \ldots.$$

Conjecture 4.1 *Let $x_0 \in \mathbb{N}$. Then every solution of Eq. (4.1) is eventually periodic with the corresponding unqiue period-3 cycle:*

$$1, \ 4, \ 2, \ 1, \ 4, \ 2, \ \ldots.$$

The upcoming example renders the enactment of the **Collatz Conjecture**.

Example 4.1 *Solve the Initial Value Problem:*

$$x_{n+1} = \begin{cases} \frac{x_n}{2} & \text{if } x_n \text{ is even,} \\ 3x_n + 1 & \text{if } x_n \text{ is odd,} \\ x_0 = 13, \end{cases} \quad n = 0, 1, \ldots,$$

and show that the solution is eventually periodic with period-3.

Solution: *Notice:*

$$x_0 = 13,$$
$$x_1 = 3 \cdot 13 + 1 = 40,$$
$$x_2 = \frac{40}{2} = 20,$$
$$x_3 = \frac{20}{2} = 10,$$
$$x_4 = \frac{10}{2} = 5,$$
$$x_5 = 3 \cdot 5 + 1 = 16,$$
$$x_6 = \frac{16}{2} = 8,$$
$$x_7 = \frac{8}{2} = 4,$$

$$x_8 = \frac{4}{2} = 2,$$

$$x_9 = \frac{2}{2} = 1.$$

We acquire seven transient terms from $(x_0 - x_6)$ *which are emphasized in square brackets,* $x_7 = x_{10}$ *and* $x_{7+n} = x_{10+n}$ *for all* $n \geq 0$:

$$[13, \ 40, \ 20, \ 10, \ 5, \ 16, \ 8], \ 4, \ 2, 1, \ \ldots.$$

We will next transition to the periodic character of the **3X+1 Conjecture** in the form:

$$x_{n+1} = \begin{cases} \frac{x_n}{2} & \text{if } x_n \text{ is even,} \\ \frac{3x_n+1}{2} & \text{if } x_n \text{ is odd,} \end{cases} \quad n = 0, 1, \ldots, \quad (4.2)$$

where $x_0 \in \mathbb{N}$. The initial condition $x_0 = 1$ generates the following unique period-2 pattern:

$$1, \ 2, \ 1, \ 2, \ \ldots.$$

Conjecture 4.2 *Let* $x_0 \in \mathbb{N}$. *Then every solution of Eq. (4.2) is eventually periodic with the following period-2 cycle:*

$$2, \ 1, \ 2, \ 1, \ \ldots.$$

The upcoming section will explore the periodic aspects of the Tent-Map.

4.2 The Tent-Map

This section will focus on the periodic traits of the **Tent-Map** in the form:

$$x_{n+1} = \begin{cases} 2x_n & \text{if } x_n < \frac{1}{2}, \\ 2(1 - x_n) & \text{if } x_n \geq \frac{1}{2}. \end{cases} \quad n = 0, 1, \ldots, \quad (4.3)$$

where $x_0 \in (0, 1)$. The upcoming examples will examine the existence, uniqueness and patterns of the period-2, period-3, and period-4 cycles of Eq. (4.3).

Example 4.2 *Determine the period-2 cycle of Eq. (4.3).*

Solution: *Set* $x_2 = x_0$ *and we acquire:*

$$x_0,$$
$$x_1 = 2x_0,$$
$$x_2 = 2(1 - x_1) = 2(1 - 2x_0) = 2 - 4x_0 = x_0.$$

Hence we obtain $x_0 = \frac{2}{5}$ *and the corresponding unique period-2 pattern (Figure 4.1):*

$$\frac{2}{5}, \frac{4}{5}, \frac{2}{5}, \frac{4}{5} \ldots \ldots \qquad (4.4)$$

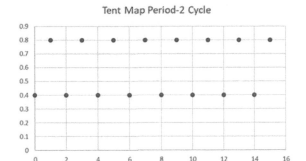

FIGURE 4.1: Tent-Map Period-2 Cycle.

Example 4.3 *Determine the period-3 cycle of Eq. (4.3).*

Solution: *Set $x_3 = x_0$ and we acquire:*

$$x_0,$$
$$x_1 = 2x_0,$$
$$x_2 = 2x_1 = 2\left[2x_0\right] = 4x_0,$$
$$x_3 = 2\left(1 - x_2\right) = 2\left(1 - 4x_0\right) = 2 - 8x_0 = x_0.$$

Thus we acquire $x_0 = \frac{2}{9}$ and the associated unique period-3 pattern:

$$\frac{2}{9}, \frac{4}{9}, \frac{8}{9}, \frac{2}{9}, \frac{4}{9}, \frac{8}{9} \ldots \ldots \qquad (4.5)$$

Notice that (4.5) renders a geometric pattern with the following sketch (Figure 4.2):

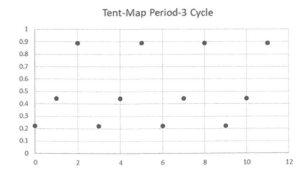

FIGURE 4.2: Tent-Map Period-3 Cycle.

Example 4.4 *Determine the period-4 cycle of Eq. (4.3).*

Solution: *Set $x_4 = x_0$ and we procure:*

$$x_0,$$
$$x_1 = 2x_0,$$
$$x_2 = 2x_1 = 2[2x_0] = 4x_0,$$
$$x_3 = 2x_2 = 2[4x_0] = 8x_0,$$
$$x_4 = 2(1-x_3) = 2(1-8x_0) = 2-16x_0 = x_0.$$

Hence we procure $x_0 = \frac{2}{17}$ and the cognate period-4 pattern:

$$\frac{2}{17}, \frac{4}{17}, \frac{8}{17}, \frac{16}{17}, \frac{2}{17}, \frac{4}{17}, \frac{8}{17}, \frac{16}{17}, \ldots \tag{4.6}$$

Observe that (4.6) emerges as a geometric pattern with the related diagram (Figure 4.3):

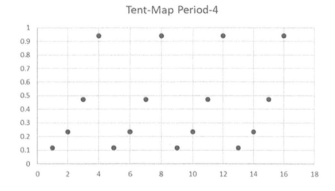

FIGURE 4.3: Tent-Map Period-4 Cycle.

Examples (4.2–4.4) guide us to the following theorem.

Theorem 4.1 *Eq. (4.3) has a unique period-p cycle ($p \geq 2$), where:*

$$x_0 = \frac{2}{2^p + 1} \tag{4.7}$$

and

$$x_i = \frac{2^{i+1}}{2^p + 1} \quad \text{for all} \quad i \in [0, 1, 2, \ldots, p-1]. \tag{4.8}$$

The proof of (4.7) and (4.8) will be left as end-of-chapter exercises. The cognate sketch (Figure 4.4) renders (4.8) as a period-7 cycle when $x_0 = \frac{2}{129}$:

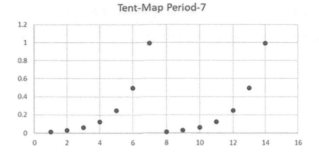

FIGURE 4.4: Tent-Map Period-7 Cycle.

Now we will transition to analyzing eventually periodic solutions of Eq. (4.3). In Chapter 1, we defined $\{x_n\}_{n=0}^{\infty}$ as an **eventually periodic sequence** with minimal period-p $(p \geq 2)$ if there exists $N \in \mathbb{N}$ such that:

$$x_{n+N} = x_{(n+p)+N} \quad \text{for all} \ \ n \geq 0. \tag{4.9}$$

Thus via (4.9), $\{x_n\}_{n=0}^{\infty}$ is eventually periodic with **minimal period-2** if there exists $N \in \mathbb{N}$ such that:

$$x_{n+N} = x_{(n+2)+N} \quad \text{for all} \ \ n \geq 0.$$

The next two examples will analytically and graphically exhibit the existence of eventually periodic solutions with N transient terms $(N \in \mathbb{N})$ of Eq. (4.3).

Example 4.5 *Solve the Initial Value Problem:*

$$
x_{n+1} = \begin{cases}
2x_n & \text{if} \ \ x_n < \frac{1}{2}\ , \\
2(1-x_n) & \text{if} \ \ x_n \geq \frac{1}{2} \\
x_0 = \frac{1}{40},
\end{cases} \quad n = 0, 1, \ldots,
$$

and show that the solution is eventually periodic with period-2.

Solution: *By iteration we acquire:*

$$\left[\frac{1}{40}, \frac{1}{20}, \frac{1}{10}, \frac{1}{5}\right], \frac{2}{5}, \frac{4}{5}, \frac{2}{5}, \frac{4}{5}, \ldots \tag{4.10}$$

*Note that the baldfaced terms in (4.10) are the four **ascending transient terms** $(x_0 - x_3)$. In addition, the transient terms $x_0 - x_3$ are resembled as the associated geometric pattern:*

$$x_i = \frac{1}{5 \cdot 2^{3-i}} \quad \text{for all} \ \ i \in [0, 1, 2, 3]. \tag{4.11}$$

Also notice that $x_4 = x_6$ and $x_{4+n} = x_{6+n}$ for all $n \geq 0$. The corresponding diagram (Figure 4.5) depicts (4.10):

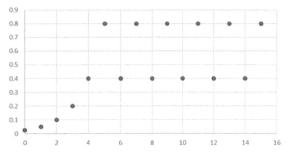

FIGURE 4.5: Eventually Periodic with Period-2 with Four Transient Terms.

Example 4.6 *Solve the Initial Value Problem:*

$$x_{n+1} = \begin{cases} 2x_n & if \quad x_n < \frac{1}{2}, \\ 2(1 - x_n) & if \quad x_n \geq \frac{1}{2} \\ x_0 = \frac{1}{288}, \end{cases} \qquad n = 0, 1, \ldots,$$

and show that the solution is eventually periodic with period-3.
Solution: *By iteration we acquire:*

$$\left[\frac{1}{288}, \frac{1}{144}, \frac{1}{72}, \frac{1}{36}, \frac{1}{18}, \frac{1}{9}\right], \frac{2}{9}, \frac{4}{9}, \frac{8}{9}, \frac{2}{9}, \frac{4}{9}, \frac{8}{9}, \ldots \qquad (4.12)$$

The baldfaced terms in (4.12) are the six **ascending transient terms**
$(x_0 - x_5)$ *that arise as a cognate geometric pattern:*

$$x_i = \frac{1}{9 \cdot 2^{5-i}} \quad \text{for all} \quad i \in [0, 1, 2, 3, 4, 5]. \qquad (4.13)$$

Observe that $x_6 = x_9$ *and* $x_{6+n} = x_{9+n}$ *for all* $n \geq 0$. *The corresponding diagram (Figure 4.6) depicts (4.12):*

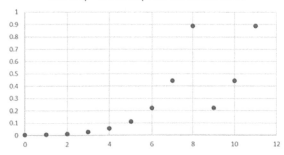

FIGURE 4.6: Eventually Periodic with Period-3 with Six Transient Terms.

Examples (4.5–4.6) direct us to the associated theorem.

Theorem 4.2 *Eq. (4.3) has an eventually periodic solution with period-p* $(p \geq 2)$ *and* $N \in \mathbb{N}$ *if:*

$$x_0 = \frac{1}{[2^p + 1] \cdot 2^{N-1}}, \tag{4.14}$$

with N $(N \in \mathbb{N})$ *corresponding geometric transient terms:*

$$x_i = \frac{1}{[2^p + 1] \cdot 2^{N-1-i}} \quad \text{for all} \quad i \in [0, 1, \ldots, N - 1]. \tag{4.15}$$

The proof of (4.14), (4.15) and Theorem (4.2) will be left as end-of-chapter exercises.

In addition to periodic and eventually periodic solutions, Eq. (4.3) also has an **eventually constant solution** where either $C = \frac{2}{3}$ or $C = 0$.

Example 4.7 *The associated sketch (Figure 4.7) describes an eventually constant solution* $(C = \frac{2}{3})$ *of Eq. (4.3) with six **ascending transient terms** with* $x_0 = \frac{1}{96}$:

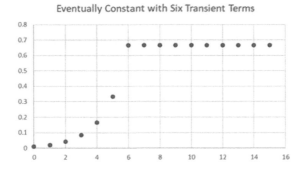

Eventually Constant with Six Transient Terms

FIGURE 4.7: Eventually Constant with Six Transient Terms.

Figure 4.7 renders the cognate pattern:

$$\left[\frac{1}{96}, \frac{1}{48}, \frac{1}{24}, \frac{1}{12}, \frac{1}{6}, \frac{1}{3} \right], \frac{2}{3}, \frac{2}{3}, \frac{2}{3}, \frac{2}{3}, \ldots \tag{4.16}$$

The six transient terms $(x_0 - x_5)$ *of (4.16) are baldfaced in square brackets and emerge as a corresponding geometric pattern:*

$$x_i = \frac{1}{3 \cdot 2^{5-i}} \quad \text{for all} \quad i \in [0, 1, 2, 3, 4, 5]. \tag{4.17}$$

In addition, $x_6 = x_7 = \frac{2}{3}$ *and* $x_{n+6} = x_{n+7} = \frac{2}{3}$ *for all* $n \geq 0$.

Example (4.7) steers us to the corresponding theorem.

Theorem 4.3 *Eq. (4.3) has an eventually constant solution* $C = \frac{2}{3}$ *if:*

$$x_0 = \frac{1}{3 \cdot 2^{N-1}}, \tag{4.18}$$

with N $(N \in \mathbb{N})$ *corresponding* **ascending transient terms***:*

$$x_i = \frac{1}{3 \cdot 2^{N-1-i}} \quad \text{for all} \quad i \in [0, 1, \ldots, N-1]. \tag{4.19}$$

The proof of (4.18), (4.19) and Theorem (4.3) will be left as end-of-chapter exercises.

Example 4.8 *The corresponding diagram (Figure 4.8) depicts an eventually constant solution $(C = 0)$ of Eq. (4.3) with eight ascending transient terms with $x_0 = \frac{1}{128}$:*

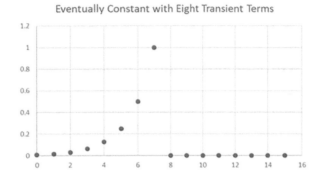

FIGURE 4.8: Eventually Constant with Eight Transient Terms.

Figure 4.8 renders the following pattern:

$$\left[\frac{1}{128}, \frac{1}{64}, \frac{1}{32}, \frac{1}{16}, \frac{1}{8}, \frac{1}{4}, \frac{1}{2}, 1 \right], 0, 0, 0, \ldots \tag{4.20}$$

The eight ascending transient terms $(x_0 - x_7)$ of (4.20) are baldfaced in square brackets and arise in the corresponding geometric pattern:

$$x_i = \frac{1}{2^{7-i}} \quad \text{for all} \quad i \in [0, 1, \ldots, 7]. \tag{4.21}$$

Furthermore, $x_8 = x_9 = 0$ and $x_{n+8} = x_{n+9} = 0$ for all $n \geq 0$.

Example (4.8) escorts us to the associated theorem.

Theorem 4.4 *Eq. (4.3) has an eventually constant solution $C = 0$ if:*

$$x_0 = \frac{1}{2^{N-1}}, \tag{4.22}$$

with N $(N \in \mathbb{N})$ *corresponding ascending transient terms:*

$$x_i = \frac{1}{2^{N-1-i}} \quad \text{for all} \quad i \in [0, 1, \ldots, N-1]. \tag{4.23}$$

The proof of (4.22), (4.23) and Theorem (4.4) will be left as end-of-chapter exercises. We will remit this section with the periodic attributes of the cognate Piece-wise Difference Equation:

$$x_{n+1} = \begin{cases} Ax_n & \text{if } x_n < \frac{1}{2}, \\ A(1 - x_n) & \text{if } x_n \geq \frac{1}{2}, \end{cases} \quad n = 0, 1, \ldots, \quad (4.24)$$

where $A > 1$ and $x_0 \in (0,1)$. Observe that Eq. (4.24) is an extension of Eq. (4.3). Determining the periodic traits of Eq. (4.24) in terms of A will be left as end-of-chapter exercises.

4.3 The Autonomous Neuron Model

This section's primary aim is to examine the existence of periodic solutions, eventually periodic solutions and eventually constant solutions of the Autonomous Piece-wise Δ.E. (discrete-time network of a single neuron model) in the form:

$$x_{n+1} = \beta x_n - g(x_n), \quad n = 0, 1, 2, \ldots, \quad (4.25)$$

where $x_0 \in \Re$, $\beta > 0$ is an internal decay rate, and g is a signal function. Eq. (4.25) is examined in several articles [2,3,6,7] and is analyzed as a single neuron model, where a signal function g is the following piece-wise constant with the **McCulloch-Pitts Nonlinearity**:

$$g(x) = \begin{cases} 1 & \text{if } x \geq 0, \\ -1 & \text{if } x < 0. \end{cases}$$

Piece-wise difference equations have been used as mathematical models for various applications, including neurons [14–18]. First of all, we will assume that

$$\beta \neq 1, \quad (4.26)$$

and we will show that Eq. (4.25) has periodic and eventually periodic solutions with period-p ($p \geq 2$). The upcoming example will depict the existence of a unique alternating period-2 cycle of Eq. (4.25).

Example 4.9 *Determine the unique alternating period-2 cycle of Eq. (4.25).*

Solution: *Assuming that $0 < x_0 < \frac{1}{\beta}$ and setting $x_2 = x_0$ we procure:*

$$x_0 > 0,$$
$$x_1 = \beta x_0 - 1,$$
$$x_2 = \beta \left[x_1\right] + 1 = \beta \left[\beta x_0 - 1\right] + 1 = \beta^2 x_0 - \beta + 1 = x_0.$$

Hence we obtain:

$$x_0 = \frac{1}{\beta + 1}, \tag{4.27}$$

and the associated alternating period-2 pattern:

$$\frac{1}{\beta + 1}, \ -\frac{1}{\beta + 1}, \ \frac{1}{\beta + 1}, \ -\frac{1}{\beta + 1}, \ \dots \tag{4.28}$$

Observe (4.28) is an alternating period-2 cycle as two neighboring terms of (4.28) are of opposite sign. The consequent diagram (Figure 4.9) renders (4.28) when $\beta = 0.25$ and $x_0 = \frac{4}{5}$:

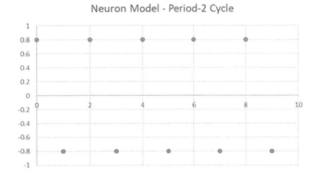

FIGURE 4.9: Autonomous Neuron Model - Alternating Period-2 Cycle.

Example 4.10 *Determine a period-3 cycle of Eq. (4.25).*

Solution: *Suppose $\frac{1}{\beta} < x_0 < \frac{\beta+1}{\beta^2}$ and $x_3 = x_0$. Then we acquire:*

$$x_0 > 0,$$
$$x_1 = \beta x_0 - 1,$$
$$x_2 = \beta [x_1] - 1 = \beta [\beta x_0 - 1] - 1 = \beta^2 x_0 - \beta - 1,$$
$$x_3 = \beta [x_2] + 1 = \beta [\beta^2 x_0 - \beta - 1] + 1 = \beta^3 x_0 - \beta^2 - \beta + 1 = x_0. \tag{4.29}$$

Hence we get:

$$x_0 = \frac{\beta^2 + \beta - 1}{\beta^3 - 1}, \tag{4.30}$$

where $\beta \neq 1$ and the related period-3 pattern:

$$\frac{\beta^2 + \beta - 1}{\beta^3 - 1}, \ \frac{\beta^2 - \beta + 1}{\beta^3 - 1}, \ \frac{-\beta^2 + \beta + 1}{\beta^3 - 1}, \ \dots \tag{4.31}$$

Note that (4.30) and (4.31) hold if and only if (4.26). The following diagram (Figure 4.10) renders (4.31) as a decreasing period-3 cycle with two positive terms and one negative term when $\beta = 3$ and $x_0 = \frac{11}{26}$:

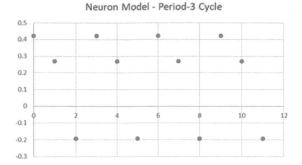

FIGURE 4.10: Autonomous Neuron Model - Decreasing Period-3 Cycle.

The period-3 cycle traced via (4.31) is not unique. For instance, by rearranging the pluses and minuses of the iterations in (4.29) we can acquire:

$$x_0 = \frac{\beta^2 - \beta - 1}{\beta^3 - 1}, \tag{4.32}$$

where $\beta \neq 1$ and the related period-3 pattern:

$$\frac{\beta^2 - \beta - 1}{\beta^3 - 1}, \ \frac{-\beta^2 - \beta + 1}{\beta^3 - 1}, \ \frac{-\beta^2 + \beta - 1}{\beta^3 - 1}, \ \dots \tag{4.33}$$

The corresponding sketch (Figure 4.11) evokes (4.33) when $\beta = 3$ and $x_0 = \frac{5}{26}$:

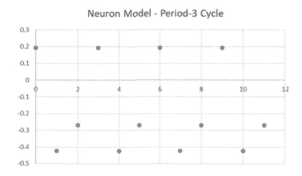

FIGURE 4.11: Autonomous Neuron Model - Period-3 Cycle.

Example 4.11 *Determine a period-4 cycle of Eq. (4.25).*

Solution: *Analogous to Example (4.10), by setting $x_4 = x_0$ we can obtain various combinations of initial conditions in the form:*

$$x_0 = \frac{\pm\beta^3 \pm \beta^2 \pm \beta \pm 1}{\beta^4 - 1} = \frac{\sum_{i=0}^{3} \pm\beta^{3-i}}{\beta^4 - 1}. \tag{4.34}$$

Not all combinations of (4.34) are possible and some of the combinations of (4.34) may reduce to a period-2 cycle. For instance, by designating

$$x_0 = \frac{\beta^3 + \beta^2 - \beta + 1}{\beta^4 - 1}, \tag{4.35}$$

when $\beta = 3$ ($x_0 = \frac{19}{40}$), we obtain the corresponding period-4 cycle with three positive terms and one negative term (Figure 4.12):

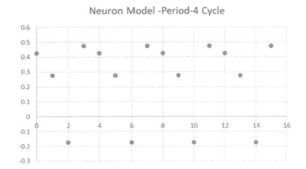

FIGURE 4.12: Autonomous Neuron Model - Period-4 Cycle.

Similarly, by selecting

$$x_0 = \frac{\beta^3 + \beta^2 - \beta - 1}{\beta^4 - 1}, \tag{4.36}$$

when $\beta = 3$ ($x_0 = \frac{19}{40}$), we acquire the cognate alternating period-4 cycle (Figure 4.13):

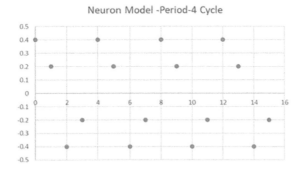

FIGURE 4.13: Autonomous Neuron Model - Alternating Period-4 Cycle.

Examples (4.10) and (4.11) then guide us to the corresponding combinations of initial conditions in determining a period-p cycle ($p \geq 3$) of Eq. (4.25):

$$x_0 = \frac{\sum_{i=0}^{p-1} \pm \beta^{p-i}}{\beta^p - 1}. \tag{4.37}$$

Now we will transition to exploring the existence and patterns of **eventually periodic solutions** of Eq. (4.25). Our objective is to determine the algorithm for describing the transient terms that give rise to eventually periodic solutions. The upcoming example will render a period-2 cycle of Eq. (4.25) with two transient terms.

Example 4.12 *Determine a period-2 solution of Eq. (4.25) with two transient terms.*

Solution: *To obtain two transient terms $(x_0 - x_1)$, we set $x_4 = x_2$ and suppose that $\dfrac{1 + \beta + \beta^2}{\beta^3} < x_0 < \dfrac{1 + \beta + \beta^2 + \beta^3}{\beta^4}$. Then by iterations we procure:*

$$x_0 > 0,$$
$$x_1 = \beta x_0 - 1,$$
$$x_2 = \beta\,[x_1] - 1 = \beta\,[\beta x_0 - 1] - 1 = \beta^2 x_0 - \beta - 1,$$
$$x_3 = \beta\,[x_2] - 1 = \beta\,[\beta^2 x_0 - \beta - 1] - 1 = \beta^3 x_0 - \beta^2 - \beta - 1, \qquad (4.38)$$
$$x_4 = \beta\,[x_3] + 1 = \beta\,[\beta^3 x_0 - \beta^2 - \beta - 1] + 1$$
$$= \beta^4 x_0 - \beta^3 - \beta^2 - \beta + 1 = x_2.$$

Via (4.42), we acquire the following initial condition:

$$x_0 = \frac{\beta^2(\beta + 1) - 2}{\beta^2(\beta^2 - 1)}, \qquad (4.39)$$

*and the corresponding pattern with two **descending transient terms** in blue in square brackets:*

$$\left[\ \frac{\beta^2(\beta + 1) - 2}{\beta^2(\beta^2 - 1)}\ ,\ \frac{\beta + 2}{\beta(\beta + 1)}\ \right],\ \frac{1}{\beta + 1}, \frac{-1}{\beta + 1}, \ \dots \qquad (4.40)$$

The upcoming sketch (Figure 4.14) traces (4.40) when $\beta = 3$ and $x_0 = \frac{17}{36}$:

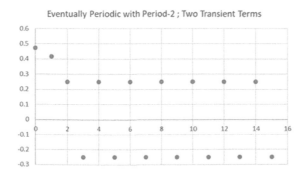

FIGURE 4.14: Eventually Periodic with Period-2 with Two Transient Terms.

Example 4.13 *Determine a period-2 solution of Eq. (4.25) with four transient terms.*

Solution: *Analogous to Example (4.12), to obtain four transient terms* $(x_0 - x_3)$, *we set* $x_6 = x_4$ *and suppose that*

$$\frac{1 + \beta + \beta^2 + \beta^3 + \beta^4}{\beta^5} < x_0 < \frac{1 + \beta + \beta^2 + \beta^3 + \beta^4 + \beta^5}{\beta^6}. \qquad (4.41)$$

Then by iterations we acquire:

$$x_0 > 0,$$
$$x_1 = \beta x_0 - 1,$$
$$x_2 = \beta [x_1] - 1 = \beta [\beta x_0 - 1] - 1 = \beta^2 x_0 - \beta - 1,$$
$$x_3 = \beta [x_2] - 1 = \beta [\beta^2 x_0 - \beta - 1] - 1 = \beta^3 x_0 - \beta^2 - \beta - 1,$$
$$x_4 = \beta [x_3] - 1 = \beta [\beta^3 x_0 - \beta^2 - \beta - 1] - 1,$$
$$= \beta^4 x_0 - \beta^3 - \beta^2 - \beta - 1, \qquad (4.42)$$
$$x_5 = \beta [x_4] - 1 = \beta [\beta^4 x_0 - \beta^3 - \beta^2 - \beta - 1] - 1$$
$$= \beta^5 x_0 - \beta^4 - \beta^3 - \beta^2 - \beta - 1,$$
$$x_6 = \beta [x_5] + 1 = \beta [\beta^5 x_0 - \beta^4 - \beta^3 - \beta^2 - \beta - 1] + 1$$
$$= \beta^6 x_0 - \beta^5 - \beta^4 - \beta^3 - \beta^2 - \beta + 1 = x_4.$$

Via (4.42), we obtain the cognate initial condition:

$$x_0 = \frac{\beta^4(\beta + 1) - 2}{\beta^4(\beta^2 - 1)}, \qquad (4.43)$$

and the associated pattern with four transient terms in blue in square brackets:

$$\left[\frac{\beta^4(\beta + 1) - 2}{\beta^4(\beta^2 - 1)} , \frac{\beta^3(\beta + 1) - 2}{\beta^3(\beta^2 - 1)} , \frac{\beta^2(\beta + 1) - 2}{\beta^2(\beta^2 - 1)} , \frac{\beta + 2}{\beta(\beta + 1)} \right],$$
$$\frac{1}{\beta + 1}, \frac{-1}{\beta + 1}, \dots \qquad (4.44)$$

The succeeding graph (Figure 4.15) depicts (4.44) when $\beta = 3$ *and* $x_0 = \frac{161}{324}$:

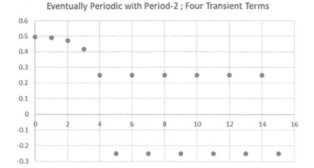

FIGURE 4.15: Eventually Periodic with Period-2 with Four Transient Terms.

Examples (4.12) and (4.13) guide us to the corresponding theorem.

Theorem 4.5 *Eq. (4.25) is eventually periodic with period-2 with N-transient terms, ($N \in \mathbb{N}$) if:*

$$x_0 = \frac{\beta^N(\beta+1) - 2}{\beta^N(\beta^2 - 1)}, \tag{4.45}$$

with the cognate pattern of the N transient terms ($x_0 - x_{N-1}$):

$$x_i = \frac{\beta^{N-i}(\beta+1) - 2}{\beta^{N-i}(\beta^2 - 1)} \quad \text{for all} \quad i \in [0, 1, 2, \ldots, N-2],$$

$$x_{N-1} = \frac{\beta+2}{\beta(\beta+1)}. \tag{4.46}$$

The proof of (4.45) extends from (4.39) and (4.43) and will be left as an end-of-chapter exercise. The proof of (4.46) follows from (4.40) and (4.44) and will also be left as an end-of-chapter exercise.

The sketch below (Figure 4.16) renders (4.45) and (4.46) with eight **descending transient terms** when $N = 8$, $\beta = 2$ and $x_0 = \frac{383}{384}$:

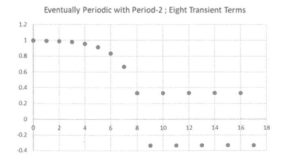

FIGURE 4.16: Eventually Periodic with Period-2 with Eight Transient Terms.

Can Eq. (4.25) exhibit an **eventually constant solution**? The answer is yes and the graph below (Figure 4.17) evokes an eventually constant solution (where $C = -1$) with eight **descending transient terms** when $\beta = 2$ and $x_0 = \frac{127}{128}$:

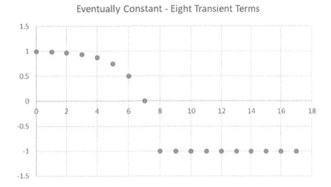

FIGURE 4.17: Eventually Constant with Eight Transient Terms.

This then leads to the following theorem.

Theorem 4.6 *Every solution of the* $\Delta.E$.

$$x_{n+1} = 2x_n - g(x_n), \quad n = 0, 1, 2, \ldots,$$

where

$$g(x) = \begin{cases} 1 & \text{if } x \geq 0, \\ -1 & \text{if } x < 0. \end{cases}$$

is eventually constant $(C = -1)$ *with N+1 transient terms,* $(N \geq 0)$ *when:*

$$x_0 = \frac{2^N - 1}{2^N}.$$

Theorem (4.6) is proved by induction and will be left as an end-of-chapter exercise. Now will transition to the periodic traits of Eq. (4.25) when $\beta = 1$.

4.3.1 Autonomous Neuron Model when $\beta = 1$

This section's aim is to explore the patterns of periodic solutions and eventually periodic solutions of Eq. (4.25) when $\beta = 1$:

$$x_{n+1} = x_n - g(x_n), \quad n = 0, 1, 2, \ldots, \tag{4.47}$$

where $x_0 \in \Re$ and

$$g(x) = \begin{cases} 1 & \text{if } x \geq 0, \\ -1 & \text{if } x < 0, \end{cases}$$

and compare the similarities and differences with the periodic attributes of Eq. (4.25) when $\beta \neq 1$. First of all, we will show that Eq. (4.47) only has periodic and eventually periodic cycles with period-2. Second of all, we will not encounter unique period-2 cycles as we did with Eq. (4.25).

Example 4.14 *Determine a period-2 cycle of Eq. (4.47).*

Solution: *Assume $0 < x_0 < 1$. Then by iterations we procure:*

$$x_0 > 0,$$
$$x_1 = x_0 - 1 < 0 \quad (as \ 0 < x_0 < 1), \qquad (4.48)$$
$$x_2 = [x_1] + 1 = [x_0 - 1] + 1 = x_0.$$

Hence via (4.48), if $0 < x_0 < 1$ then every solution of Eq. (4.47) is periodic with the associated period-2 pattern:

$$x_0, \ x_0 - 1, \ x_0, \ x_0 - 1, \ldots \qquad (4.49)$$

Via (4.49), if $x_0 = 0$ then we procure:

$$0, \ -1, \ 0, \ -1, \ \ldots.$$

Analogous to (4.48) when $-1 < x_0 < 0$, every solution of Eq. (4.47) is periodic with the corresponding period-2 pattern:

$$x_0, \ x_0 + 1, \ x_0, \ x_0 + 1, \ldots \qquad (4.50)$$

Thus via (4.48), (4.49) and (4.50) if either $x_0 \geq 1$ or $x_0 < -1$, then every solution of Eq. (4.47) is eventually periodic with period-2.

Now we will focus on determining eventually periodic solutions (transient terms) with period-2 of Eq. (4.47).

Example 4.15 *Determine the pattern of eventually periodic solution with period-2 of Eq. (4.47) when $x_0 = 4$.*

Solution: *By iterations, we procure:*

$$[4, \ 3, \ 2, \ 1], \ 0, \ -1, \ 0, \ -1, \ \ldots \qquad (4.51)$$

Note that (4.51) has four **descending transient linear terms** *$(x_0 - x_3)$ in square brackets and $x_{4+n} = x_{6+n}$ for all $n \geq 0$ resembled with the cognate sketch (Figure 4.18) below:*

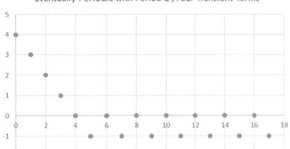

FIGURE 4.18: Eventually Periodic with Period-2 with Four Transient Terms.

Via (4.51) and Figure 4.18, we can see that the four transient terms decrease linearly with increments of 1.

On the contrary, when $x_0 = -4$, the associated sketch (Figure 4.19) renders the cognate eventually periodic solution with period-2:

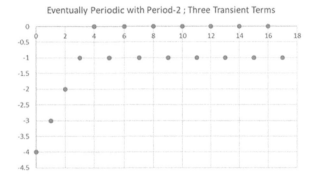

FIGURE 4.19: Eventually Periodic with Period-2 with Three Transient Terms.

In comparison to Figure 4.18, Figure 4.19 renders the associated pattern:

$$[-4, \ -3, \ -2], \ -1, \ 0, \ -1, \ 0, \ \ldots, \tag{4.52}$$

with three **ascending transient terms** *$(x_0 - x_2)$ in square brackets and $x_{3+n} = x_{5+n}$ for all $n \geq 0$.*

From Example (4.15), via (4.51) and Figure 4.18, it follows that if $x_0 = N$ ($N \in \mathbb{N}$), then Eq. (4.47) is eventually periodic with period-2 with the corresponding N transient terms:

$$x_i = N - i \quad \text{for all} \quad i \in [0, 1, 2, \ldots, N-1]. \tag{4.53}$$

Analogous to (4.51), the N transient terms in (4.53) decrease linearly in increments of 1. On the other hand, if $x_0 = -N$ where (N $\geq$ 2), then Eq. (4.47) is eventually periodic with period-2 with the associated $N-1$ transient terms:

$$x_i = -N + i \quad \text{for all} \quad i \in [0, 1, 2, \ldots, N-2]. \tag{4.54}$$

Similar to (4.52), the $N-1$ transient terms in (4.54) increase linearly in increments of 1. The transient terms of Eq. (4.47) either increase or decrease linearly in comparison to the pattern of the transient terms in the previous section when $\beta \neq 1$.

4.4   Non-Autonomous Neuron Model

This section's aim is to explore the periodic attributes of the Non-Autonomous Piece-wise Δ.E.:

$$x_{n+1} = \beta_n x_n - g(x_n), \quad n = 0, 1, 2, \ldots, \tag{4.55}$$

where $x_0 \in \Re$, $\{\beta_n\}_{n=0}^{\infty}$ is a period-2 sequence and

$$g(x) = \begin{cases} 1 & \text{if } x \geq 0, \\ -1 & \text{if } x < 0. \end{cases}$$

We will show that Eq. (4.55) only has even-ordered periodic cycles with period-2k, $(k \in \mathbb{N})$ provided that $\beta_0 \beta_1 \neq 1$. The succeeding examples will ascertain period-2 cycles and period-4 cycles of Eq. (4.55) when $\beta_0 \beta_1 \neq 1$.

Example 4.16 *Determine the period-2 cycle of Eq. (4.55).*

Solution: *Suppose that $0 < x_0 < \frac{1}{\beta_0}$ and set $x_2 = x_0$. Then we obtain:*

$$\begin{aligned} &x_0 > 0, \\ &x_1 = \beta_0 x_0 - 1, \\ &x_2 = \beta_1 [x_1] + 1 = \beta_1 [\beta_0 x_0 - 1] + 1 = \beta_0 \beta_1 x_0 - \beta_1 + 1 = x_0. \end{aligned} \tag{4.56}$$

Via (4.56), we acquire:

$$x_0 = \frac{1 - \beta_1}{\beta_0 \beta_1 - 1} \tag{4.57}$$

where $\beta_0 \beta_1 \neq 1$ and the cognate period-2 pattern:

$$\frac{1 - \beta_1}{\beta_0 \beta_1 - 1}, \frac{\beta_0 - 1}{\beta_0 \beta_1 - 1}, \ldots \tag{4.58}$$

The upcoming sketch (Figure 4.20) renders (4.58) as a decreasing period-2 pattern when instead of with $\beta_0 = 2$, $\beta_1 = 3$ and $x_0 = 0.4$:

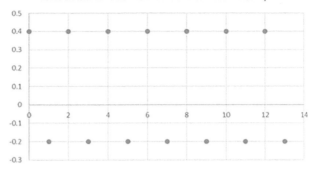

FIGURE 4.20: Non-autonomous Neuron Model - Period-2 Cycle.

Example 4.17 *Determine the period-4 cycle of Eq. (4.55).*

Solution: *Set $x_4 = x_0$ and assume that $0 < x_0 < \frac{1}{\beta_0}$. By iteration we acquire:*

$$
\begin{aligned}
x_0 &> 0, \\
x_1 &= \beta_0 x_0 - 1, \\
x_2 &= \beta_1 [x_1] + 1 = \beta_1 [\beta_0 x_0 - 1] + 1 = \beta_0 \beta_1 x_0 - \beta_1 + 1, \\
x_3 &= \beta_0 [x_2] + 1 = \beta_0 [\beta_0 \beta_1 x_0 - \beta_1 + 1] + 1 \\
&= \beta_0^2 \beta_1 x_0 - \beta_0 \beta_1 + \beta_0 + 1, \\
x_4 &= \beta_1 [x_3] - 1 = \beta_1 [\beta_0^2 \beta_1 x_0 - \beta_0 \beta_1 + \beta_0 + 1] - 1 \\
&= \beta_0^2 \beta_1^2 x_0 - \beta_0 \beta_1^2 + \beta_0 \beta_1 - 1 = x_0.
\end{aligned}
\tag{4.59}
$$

Via (4.59), we procure:

$$
x_0 = \frac{\beta_1 - 1}{\beta_0 \beta_1 + 1}
\tag{4.60}
$$

where $\beta_0 \beta_1 \neq -1$ and the associated alternating period-4 pattern:

$$
\frac{\beta_1 - 1}{\beta_0 \beta_1 + 1}, \quad \frac{-(\beta_0 + 1)}{\beta_0 \beta_1 + 1}, \quad \frac{1 - \beta_1}{\beta_0 \beta_1 + 1}, \quad \frac{\beta_0 + 1}{\beta_0 \beta_1 + 1}, \quad \dots.
\tag{4.61}
$$

The succeeding diagram (Figure 4.21) depicts (4.61) with $\beta_0 = 1$, $\beta_1 = 4$ and $x_0 = 0.6$:

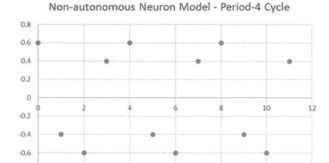

FIGURE 4.21: Non-autonomous Neuron Model - Period-4 Cycle.

From Examples (4.16) and (4.17), to obtain a period-6 cycle of Eq. (4.55), we set $x_6 = x_0$ and apply analogous techniques in (4.56) and (4.59) and procure:

$$x_0 = \frac{\beta_0^3 \beta_1^2 + \beta_0^2 \beta_1^2 - \beta_0 \beta_1^2 - \beta_0 \beta_1 + \beta_1 - 1}{1 - (\beta_0 \beta_1)^3}. \tag{4.62}$$

Using (4.62), the sketch (Figure 4.22) below renders a period-6 cycle of Eq. (4.55) when $\beta_0 = 1$, $\beta_1 = 2$ and $x_0 = -\frac{5}{7}$:

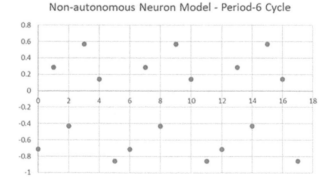

FIGURE 4.22: Non-autonomous Neuron Model - Period-6 Cycle.

Notice that in Figures 4.20 and 4.22, the period-2 and period-6 cycles are not alternating cycles. On the other hand, Figure 4.21 renders an alternating period-4 cycle. The next graph (Figure 4.23) renders an alternating period-8 cycle when $\beta_0 = 1$, $\beta_1 = 3$ and $x_0 = 0.4$:

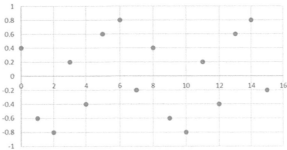

FIGURE 4.23: Non-autonomous Neuron Model - Period-8 Cycle.

From Figures 4.21 and 4.23, can we conclude that period-4k cycles ($k \in \mathbb{N}$) of Eq. (4.55) are alternating periodic cycles? From Figures 4.20 and 4.22, can we conclude that period-4k+2 cycles ($k \geq 0$) of Eq. (4.55) are non-alternating periodic cycles? This will be left to examine as end-of-chapter exercises. Now we will shift our focus on examining an eventually periodic cycle with period-2 of Eq. (4.55) with 2N transient terms, ($N \in \mathbb{N}$). We will show that the transient terms of Eq. (4.55) will decrease in two subsequences versus monotonic decrease encountered in Examples (4.12) and (4.13) of Eq. (4.25).

The succeeding examples will mimic the tactics implemented in Examples (4.12) and (4.13) to determine the pattern of the transient terms of Eq. (4.25) expressed in terms of β_0 and β_1.

Example 4.18 *Determine the period-2 cycle with two transient terms of Eq. (4.55).*
Solution: *Analogous to Example (4.12), to acquire the two transient terms $(x_0 - x_1)$, we set $(x_4 = x_2)$ and assume that*

$$\frac{\beta_1 + 1}{\beta_0 \beta_1} < x_0 < \frac{\beta_0 [\beta_1 + 1] + 1}{\beta_0 [\beta_0 \beta_1]}. \tag{4.63}$$

Then by iteration we procure:

$$x_0 > 0,$$
$$x_1 = \beta_0 x_0 - 1,$$
$$x_2 = \beta_1 [x_1] - 1 = \beta_1 [\beta_0 x_0 - 1] - 1 = \beta_0 \beta_1 x_0 - \beta_1 - 1,$$
$$x_3 = \beta_0 [x_2] - 1 = \beta_0 [\beta_0 \beta_1 x_0 - \beta_1 - 1] - 1 \tag{4.64}$$
$$\qquad = \beta_0^2 \beta_1 x_0 - \beta_0 \beta_1 - \beta_0 - 1,$$
$$x_4 = \beta_1 [x_3] + 1 = \beta_1 [\beta_0^2 \beta_1 x_0 - \beta_0 \beta_1 - \beta_0 - 1] + 1$$
$$\qquad = \beta_0^2 \beta_1^2 x_0 - \beta_0 \beta_1^2 - \beta_0 \beta_1 - \beta_1 + 1 = x_2.$$

Via (4.64), we obtain the following initial condition:

$$x_0 = \frac{(\beta_0\beta_1)[\beta_1 + 1] - 2}{(\beta_0\beta_1)[\beta_0\beta_1 - 1]}, \tag{4.65}$$

where $\beta_0\beta_1 \neq 1$ and the cognate pattern of the transient terms (x_0 in blue and x_1 in green) in square brackets:

$$\left[\frac{(\beta_0\beta_1)[\beta_1 + 1] - 2}{(\beta_0\beta_1)[\beta_0\beta_1 - 1]} , \frac{\beta_1[\beta_0 + 1] - 2}{\beta_1[\beta_0\beta_1 - 1]} \right], \frac{\beta_1 - 1}{\beta_0\beta_1 - 1}, \frac{1 - \beta_0}{\beta_0\beta_1 - 1}, \cdots \tag{4.66}$$

Note that (4.66) shares similar characteristics with (4.40) in Example (4.12) where $x_{n+2} = x_n$ for all $n \geq 2$. The consequent sketch (Figure 4.24) traces (4.66) when $\beta_0 = 2$, $\beta_1 = 3$ and $x_0 = \frac{11}{15}$:

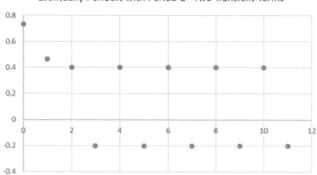

FIGURE 4.24: Eventually Periodic with Period-2 with Two Transient Terms.

Example 4.19 *Determine the period-2 cycle with four transient terms of Eq. (4.55).*

Solution: *Similar to Example (4.13), to acquire the four transient terms $x_0 - x_3$ assume that*

$$\frac{[\beta_1 + 1][\beta_0\beta_1 + 1]}{[\beta_0\beta_1]^2} < x_0 < \frac{\beta_0[\beta_1 + 1][\beta_0\beta_1 + 1] + 1}{\beta_0[\beta_0\beta_1]^2}. \tag{4.67}$$

Parallel to the techniques in Examples (4.12), (4.13) and (4.18) by setting $x_6 = x_4$, we acquire the corresponding initial condition:

$$x_0 = \frac{(\beta_0\beta_1)^2[\beta_1 + 1] - 2}{(\beta_0\beta_1)^2[\beta_0\beta_1 - 1]}, \tag{4.68}$$

where $\beta_0\beta_1 \neq 1$ and the related pattern of the four transient terms $(x_0 - x_3)$:

$$\frac{(\beta_0\beta_1)^2 [\beta_1 + 1] - 2}{(\beta_0\beta_1)^2 [\beta_0\beta_1 - 1]} \; , \; \frac{\beta_1(\beta_0\beta_1)[\beta_0 + 1] - 2}{\beta_1(\beta_0\beta_1)[\beta_0\beta_1 - 1]} \; , \; \frac{\beta_0\beta_1[\beta_1 + 1] - 2}{\beta_0\beta_1[\beta_0\beta_1 - 1]} \; ,$$

$$\frac{\beta_1[\beta_0 + 1] - 2}{\beta_1[\beta_0\beta_1 - 1]} \; . \tag{4.69}$$

To formulate (4.69), we will decompose into an even-ordered sub-sequence in blue and an odd-ordered sub-sequence in green and obtain:

$$x_i = \begin{cases} \dfrac{(\beta_0\beta_1)^{\frac{4-i}{2}} [\beta_1+1] - 2}{(\beta_0\beta_1)^{\frac{4-i}{2}} [\beta_0\beta_1 - 1]} & \text{if} \quad i \in [0, 2] \; , \\[4mm] \dfrac{\beta_1(\beta_0\beta_1)^{\frac{3-i}{2}} [\beta_0+1] - 2}{\beta_1(\beta_0\beta_1)^{\frac{3-i}{2}} [\beta_0\beta_1 - 1]} & \text{if} \quad i \in [1, 3] \; . \end{cases} \tag{4.70}$$

The subsequent representation (Figure 4.25) depicts (4.68), (4.69), and (4.70) when $\beta_0 = 2$, $\beta_1 = 3$ and $x_0 = \frac{71}{90}$:

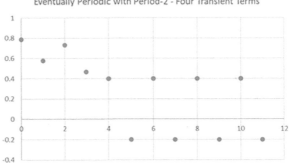

FIGURE 4.25: Eventually Periodic with Period-2 with Four Transient Terms.

Examples (4.18) and (4.19) guide us to the resultant theorem.

Theorem 4.7 *Eq. (4.55) is eventually periodic with period-2 with 2N transient terms $(N \in \mathbb{N})$ if:*

$$x_0 = \frac{(\beta_0\beta_1)^N [\beta_1 + 1] - 2}{(\beta_0\beta_1)^N [\beta_0\beta_1 - 1]}, \tag{4.71}$$

with the associated pattern of the 2N transient terms $(x_0 - x_{2N-1})$:

$$x_i = \begin{cases} \dfrac{(\beta_0\beta_1)^{\frac{2N-i}{2}} [\beta_1+1] - 2}{(\beta_0\beta_1)^{\frac{2N-i}{2}} [\beta_0\beta_1 - 1]} & \text{if} \quad i \in [0, 2, 4, \ldots, 2N-2] \; , \\[4mm] \dfrac{\beta_1(\beta_0\beta_1)^{\frac{2N-1-i}{2}} [\beta_0+1] - 2}{\beta_1(\beta_0\beta_1)^{\frac{2N-1-i}{2}} [\beta_0\beta_1 - 1]} & \text{if} \quad i \in [1, 3, 5, \ldots, 2N-1] \; . \end{cases} \tag{4.72}$$

The proof of (4.71) advances from (4.65) and (4.68) and will be left as an end-of-chapter exercise. The proof of (4.72) proceeds from (4.66), (4.69) and (4.70) and will also be left as an end-of-chapter exercise. The sketch below (Figure 4.26) renders (4.71), and (4.72) when $N = 6$, $\beta_0 = \frac{3}{2}$, $\beta_1 = 2$ and $x_0 = \frac{2185}{1458}$:

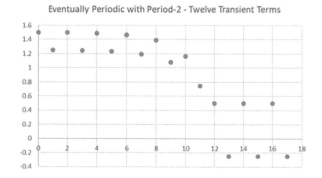

FIGURE 4.26: Eventually Periodic with Period-2 with Twelve Transient Terms.

4.4.1   Non-Autonomous Neuron Model when $\beta_0\beta_1 = 1$

This section's objective is to detect the patterns of periodic solutions and eventually periodic solutions of Eq. (4.55) when $\beta_0\beta_1 = 1$:

$$x_{n+1} = \beta_n x_n - g(x_n), \quad n = 0, 1, 2, \ldots, \tag{4.73}$$

where $x_0 \in \Re$, $\{\beta_n\}_{n=0}^{\infty}$ is a period-2 sequence, $\beta_0\beta_1 = 1$ and

$$g(x) = \begin{cases} 1 & \text{if } x \geq 0, \\ -1 & \text{if } x < 0, \end{cases}$$

and compare the similarities and differences with the periodic properties of Eq. (4.55) when $\beta_0\beta_1 \neq 1$. We will show that Eq. (4.73) will only exhibit periodic and eventually periodic cycles with period-4 that are not unique, in contrast with Eq. (4.55) that portrays periodic or eventually periodic cycles with period-2k ($k \in \mathbb{N}$).

Example 4.20 *Determine a period-4 cycle of Eq. (4.73).*

Solution: *Since we assumed that $\beta_0\beta_1 = 1$, then either $\beta_0 > 1$ and $\beta_1 < 1$ or vice versa, which then leads to the two corresponding cases:*

- **Case 1**: *The period-4 pattern when $\beta_0 > 1$, $\beta_1 < 1$ and $x_0 = -1$:*

$$-1, \ 1 - \beta_0, \ \beta_1, \ 0, \ \ldots. \tag{4.74}$$

- **Case 2**: *The period-4 pattern when* $\beta_0 < 1$, $\beta_1 > 1$ *and* $x_0 = 0$:

$$0, \ -1, \ 1 - \beta_1, \ \beta_0, \ \ldots. \tag{4.75}$$

Cases 1 and 2 render different patterns. The upcoming diagram (Figure 4.27) renders a non-zero period-4 cycle when $\beta_0 = 2$, $\beta_1 = 0.5$ *and* $x_0 = 0.6$:

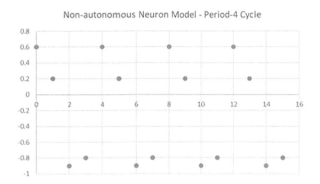

FIGURE 4.27: Non-autonomous Neuron Model - Period-4 Cycle.

We can also show that (4.74) and (4.75) are not the unique period-4 cycles of Eq. (4.73) and can exhibit non-zero period-4 cycles as we can see in Figure (4.27). This will be left as an end-of-chapter exercise.

Now we will transition to examining the patterns of eventually periodic cycles with period-4 of Eq. (4.73). The consequent diagram (Figure 4.28) renders an eventually periodic solution with period-4 of Eq. (4.73) with six transient terms decomposed as two decreasing sub-sequences when $\beta_0 = 2$, $\beta_1 = 0.5$ and $x_0 = 5$:

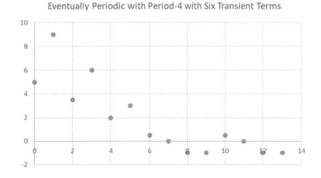

FIGURE 4.28: Eventually Periodic with Period-4 with Six Transient Terms.

The upcoming example will outline a method of determining the transient behavior of periodic cycles with period-4 of Eq. (4.73).

Example 4.21 *Determine an eventually periodic cycle with period-4 of Eq. (4.73).*

Solution: *Suppose:*

(1) $\beta_0\beta_1 = 1$.

(2) $\beta_0 < 1$ *and* $\beta_1 > 1$.

(3) $\frac{4}{3} < \beta_1 < \frac{3}{2}$.

(4) $x_0 = \beta_1 + 2$.

Then by iterations we acquire:

$$
\begin{aligned}
x_0 &= \beta_1 + 2, \\
x_1 &= \beta_0\,[x_0] - 1 = \beta_0\beta_1 + 2\beta_0 - 1 = 2\beta_0, \\
x_2 &= \beta_1\,[x_1] - 1 = 2\beta_0\beta_1 - 1 = 1, \\
x_3 &= \beta_0\,[x_2] - 1 = \beta_0 - 1 < 0, \\
x_4 &= \beta_1\,[x_3] + 1 = \beta_0\beta_1 - \beta_1 + 1 = 2 - \beta_1 > 0, \\
x_5 &= \beta_0\,[x_4] - 1 = 2\beta_0 - \beta_0\beta_1 - 1 = 2\beta_0 - 2 = 2\,(\beta_0 - 1) < 0, \\
x_6 &= \beta_1\,[x_5] + 1 = 2\beta_0\beta_1 - 2\beta_1 + 1 = 3 - 2\beta_1 > 0, \\
x_7 &= \beta_0\,[x_6] - 1 = 3\beta_0 - 2\beta_0\beta_1 - 1 = 3\,(\beta_0 - 1) < 0, \\
x_8 &= \beta_1\,[x_7] + 1 = 3\beta_0\beta_1 - 3\beta_1 + 1 = 4 - 3\beta_1 > 0, \\
x_9 &= \beta_0\,[x_6] + 1 = 4\beta_0 - 3\beta_0\beta_1 + 1 = 4\beta_0 - 2 > 0, \\
x_{10} &= \beta_1\,[x_7] - 1 = 4\beta_0\beta_1 - 2\beta_1 - 1 = 3 - 2\beta_1 = x_6.
\end{aligned}
\tag{4.76}
$$

Note that (4.76) renders six transient terms $(x_0 - x_5)$. Furthermore, we can conclude that there exists $k \in \mathbb{N}$ such that:

$$
\frac{k+2}{k+1} \le \beta_1 \le \frac{k+1}{k}.
$$

as shown in Figure 4.29.

FIGURE 4.29: Choosing $\beta_1 \in \left[\frac{k+2}{k+1}, \frac{k+1}{k}\right]$ for some $k \in \mathbb{N}$.

Now notice that if $\beta_1 = \frac{k+1}{k}$ for some $k \in \mathbb{N}$, then Eq. (4.73) renders the corresponding 2(k+1) transient terms $(x_0 - x_{2k+1})$:

$$x_i = \begin{cases} \beta_1 + 2 & \text{if } i = 0, \\ 2\beta_0 & \text{if } i = 1, \\ 1 & \text{if } i = 2, \\ [j+1](\beta_0 - 1) & \text{if } i = 3 + 2j \text{ for all } j \in [0, 1, \ldots, k], \\ (m+1) - (m)\beta_1 & \text{if } i = 4 + 2m \text{ for all } m \in [0, 1, \ldots, k-1], \end{cases} \tag{4.77}$$

and for all $n \geq 0$ the corresponding period-4 pattern:

$$x_{2k+2+4n} = 0, \; x_{2k+3+4n} = -1, \; x_{2k+4+4n} = 1 - \beta_1, \; x_{2k+5+4n} = \beta_0. \tag{4.78}$$

Observe that (4.77) extends from (4.76) and will be left as end-of-chapter exercises to prove. In addition, (4.78) shows exactly the same pattern as in (4.75). The sketch below (Figure 4.30) renders (4.77) and (4.78) as an eventually periodic cycle with period-4 with ten transient terms $(x_0 - x_9)$ when $k = 4$, $\beta_0 = \frac{4}{5}$, $\beta_1 = \frac{5}{4}$, and $x_0 = \beta_1 + 2 = 3.25$:

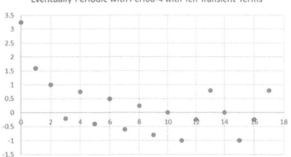

Eventually Periodic with Period-4 with Ten Transient Terms

FIGURE 4.30: Eventually Periodic with Period-4 with Ten Transient Terms.

On the other hand, if $\beta_1 \neq \frac{k+1}{k}$ for all $k \in \mathbb{N}$, then Eq. (4.73) describes the cognate 2k transient terms $(x_0 - x_{2k-1})$:

$$x_i = \begin{cases} \beta_1 + 2 & \text{if } i = 0, \\ 2\beta_0 & \text{if } i = 1, \\ 1 & \text{if } i = 2, \\ [j+1](\beta_0 - 1) & \text{if } i = 3 + 2j \text{ for all } j \in [0, 1, \ldots, k-1], \\ (m+2) - (m+1)\beta_1 & \text{if } i = 4 + 2m \text{ for all } m \in [0, 1, \ldots, k-2], \end{cases} \tag{4.79}$$

and for all $n \geq 0$ and the associated period-4 pattern:

$$\begin{cases} x_{2k+4n} = k - (k-1)\beta_1, \\ x_{2k+1+4n} = k(\beta_0 - 1), \\ x_{2k+2+4n} = (k+1) - k\beta_1, \\ x_{2k+3+4n} = (k+1)\beta_0 - (k-1), \end{cases} \tag{4.80}$$

Notice that (4.80) is not a unique period-4 cycle as its pattern depends on k, in comparison to the unique period-4 cycle in (4.78). In addition, (4.77) has $2k + 2$ transient terms, but (4.79) has $2k$ transient terms, even though they depict very similar structures.

Furthermore, (4.79) and (4.80) extend from (4.76) and will be left as end-of-chapter exercises to prove. The corresponding diagram (Figure 4.31) depicts (4.79) and (4.80) as an eventually periodic cycle with period-4 with eight transient terms $(x_0 - x_7)$ when $k = 4$, $\beta_0 = \frac{16}{25}$, $\beta_1 = \frac{25}{16}$, and $x_0 = \beta_1 + 2 = 3.28$:

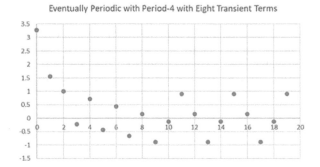

FIGURE 4.31: Eventually Periodic with Period-4 with Eight Transient Terms.

More studies on Piece-wise Difference Equations will be addressed with the corresponding questions:

- When $\{\beta_n\}_{n=0}^{\infty}$ is a period-3 sequence:

    - Will Eq. (4.55) have periodic and eventually periodic cycles with only period-3N, $(N \geq 1)$?
    - How many transient terms will Eq. (4.55) exhibit?
    - Periodic and eventually periodic cycles of Eq. (4.55) when $\beta_0 \beta_1 \beta_2 = 1$?

- Periodic and eventually periodic cycles of Eq. (4.55) when $\{\beta_n\}_{n=0}^{\infty}$ is a period-k sequence, $(k \geq 4)$?

We will adjourn this chapter with additional models used as Piece-wise Difference Equations.

4.5   The Williamson Model

The **Williamson Model** established by M. Williamson in 1974 [13] who studied age-structured cases in a discrete population model resembled by the following Piece-wise Δ.E.:

$$x_{n+1} = \begin{cases} \lambda_+ x_n & \text{if } x_n > 1 \,, \\ \lambda_- x_n & \text{if } x_n < 1 \,, \end{cases} \quad n = 0, 1, \dots,$$

where $x_0 > 0$, $0 < \lambda_- < 1 < \lambda_+$, and λ_+ and λ_- are the population growth rates.

4.6 The West Nile Epidemics Model

The **West Nile Epidemics Model** established by Vlajko Kocic in 2006 [9] is resembled by the following Piece-wise Δ.E.:

$$x_{n+1} = [a - bh(x_n - c)]\, x_n, \quad n = 0, 1, \dots,$$

where h is the heaviside function defined as:

$$h(t) = \begin{cases} 0 & \text{if } t < 0 \,, \\ 1 & \text{if } t \geq 0 \,, \end{cases}$$

where $x_0 \geq 0$, $0 < b < 1 < a < b + 1$, $a - b \in (0,1)$ and $c > 0$. This Piece-wise Δ.E. appears in the discrete model of the West Nile Epidemics when spraying against mosquitoes is implemented only when the number of mosquitoes exceeds some predefined threshold level.

4.7 Chapter 4 Exercises

Consider the **Tent-Map** in the form:

$$x_{n+1} = \begin{cases} 2x_n & \text{if } x_n < \frac{1}{2} \,, \\ 2(1 - x_n) & \text{if } x_n \geq \frac{1}{2} \,. \end{cases} \quad n = 0, 1, \dots,$$

where $0 < x_0 < 1$. In problems 1–14:

1. Determine a period-3 pattern with two transient terms.

2. Determine a period-3 pattern with three transient terms.

3. Determine a period-3 pattern with four transient terms.

4. From Exercises 1–3, determine a period-3 pattern with N transient terms ($N \in \mathbb{N}$).

5. Determine a period-4 pattern with two transient terms.

6. Determine a period-4 pattern with three transient terms.

7. Determine a period-4 pattern with four transient terms.

8. From Exercises 5–8, determine a period-4 pattern with N transient terms $(N \in \mathbb{N})$.

9. Determine a period-p $(p \geq 2)$ pattern with two transient terms.

10. Determine a period-p $(p \geq 2)$ pattern with three transient terms.

11. Determine a period-p $(p \geq 2)$ pattern with four transient terms.

12. From Exercises 9–11, determine a period-p $(p \geq 2)$ pattern with N transient terms $(N \in \mathbb{N})$.

Consider the **Tent-Map** in the form:

$$x_{n+1} = \begin{cases} Ax_n & \text{if } x_n < \frac{1}{2}, \\ A(1 - x_n) & \text{if } x_n \geq \frac{1}{2}. \end{cases} \qquad n = 0, 1, \ldots,$$

where $0 < x_0 < 1$ and $A > 2$. In problems 13–20:

13. Determine a period-2 pattern.

14. Determine a period-3 pattern.

15. Determine a period-4 pattern.

16. From Exercises 13–15, determine a period-p pattern $(p \geq 2)$.

17. Determine a period-2 pattern with two transient terms.

18. Determine a period-3 pattern with four transient terms.

19. Determine a period-4 pattern with six transient terms.

20. From Exercises 17–19, determine a period-k pattern (for $k \geq 2$) with N transient terms $(N \in \mathbb{N})$.

Consider the **Piece-wise** Δ.**E.** in the form:

$$x_{n+1} = \beta x_n - g(x_n), \quad n = 0, 1, 2, \ldots,$$

where $\beta > 0$ and

$$g(x) = \begin{cases} 1, & x \geq 0, \\ -1, & x < 0. \end{cases}$$

In problems 21–32:

21. Determine a period-3 pattern.

22. Determine a period-4 pattern.

23. Determine a period-5 pattern.

24. From Exercises 21–23, determine a period-p pattern $(p \geq 3)$.

25. Determine a period-3 pattern with two transient terms.

26. Determine a period-3 pattern with three transient terms.

27. Determine a period-3 pattern with four transient terms.

28. From Exercises $25 - 27$, determine a determine a period-3 pattern with N transient terms ($N \in \mathbb{N}$).

29. Determine a period-4 pattern with two transient terms.

30. Determine a period-4 pattern with three transient terms.

31. Determine a period-4 pattern with four transient terms.

32. From Exercises 29–31, determine a determine a period-4 pattern with N transient terms ($N \in \mathbb{N}$).

Chapter 5

Max-Type Difference Equations

This chapter's aim is to examine the patterns of periodic and eventually periodic solutions of the corresponding second-order Autonomous Max-Type Δ.E.:

$$x_{n+2} = \max\left\{\frac{A}{x_{n+1}}, \frac{B}{x_n}\right\}, \ n = 0, 1, \ldots, \tag{5.1}$$

where $A, B, x_0, x_1 > 0$ and the second-order Non-Autonomous Max-Type Δ.E.:

$$x_{n+2} = \max\left\{\frac{1}{x_{n+1}}, \frac{C_n}{x_n}\right\}, \ n = 0, 1, \ldots, \tag{5.2}$$

where $x_0, x_1 > 0$ and $\{C_n\}_{n=0}^{\infty}$ is a periodic sequence of positive real numbers with period-k $(k \geq 2)$. Observe that Eq. (5.1) is assembled as combinations of the first- and second-order Autonomous Riccati Difference Equations, while Eq. (5.2) is composed as combinations of the Autonomous first-order Riccati Difference Equation and the Non-Autonomous second-order Riccati Difference Equation.

Our primary aim throughout this chapter is to determine the patterns of periodic cycles and the patterns of the transient terms of Eqs. (5.1) and (5.2). Analogous to Eqs. (4.55) and (4.73), the transient terms of Eqs. (5.1) and (5.2) will be devised as piece-wise sequences.

5.1   The Autonomous Case (Eq. [5.1])

This section's aim is to meticulously examine the periodic traits of Eq. (5.1). By a change of variables, we can reformulate (5.1) as:

$$x_{n+2} = \max\left\{\frac{1}{x_{n+1}}, \frac{C}{x_n}\right\}, \ n = 0, 1, \ldots, \tag{5.3}$$

where $C, x_0, x_1 > 0$. Notice that the left side of Eq. (5.3) is the first-order Autonomous Riccati Difference Equation (Eq. (3.3)), while the right side of Eq. (5.3) is the second-order Autonomous Riccati Difference Equation (Eq. (3.15)).

In [1], it was shown that every positive solution of Eq. (5.3) is eventually periodic with the associated periods:

$$
\begin{cases}
2 & \text{if } C < 1, \\
3 & \text{if } C = 1, \\
4 & \text{if } C > 1.
\end{cases}
$$

We will show that the periodic cycles of Eq. (5.3) are not unique and will depend on the initial conditions x_0 and x_1 and the powers of C (when $C \neq 1$ for period-2 and period-4 cycles). The transient terms of Eq. (5.3) will be contrived as three piece-wise sequences with powers of C (when $C \neq 1$ for period-2 and period-4 cycles).

Now we will examine some graphical examples of periodic and eventually periodic solutions of Eq. (5.3). For instance, the corresponding sketch (Figure 5.1) renders and an eventually periodic cycle with period-2 with fourteen transient terms when $C = 0.8$, $x_0 = 0.3125$ and $x_1 = 3.2$ (note that x_0 and x_1 are reciprocals):

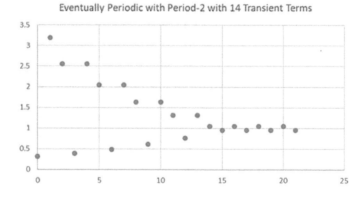

Eventually Periodic with Period-2 with 14 Transient Terms

FIGURE 5.1: Eventually Periodic with Period-2 with Fourteen Transient Terms.

The upcoming diagram (Figure 5.2) depicts an eventually periodic cycle with period-4 with twelve transient terms when $C = 2$, $x_0 = 16$ and $x_1 = 0.0625$ (note that x_0 and x_1 are reciprocals):

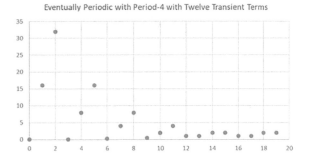

FIGURE 5.2: Eventually Periodic with Period-4 with Twelve Transient Terms.

Next we will render examples of assorted period-3 cycles of Eq. (5.3) when $C = 1$. The upcoming sketch (Figure 5.3) renders a **descending triangular-shaped** period-3 cycle when $C = 1$, $x_0 = 2.5$ and $x_1 = 0.4$ (note that x_0 and x_1 are reciprocals):

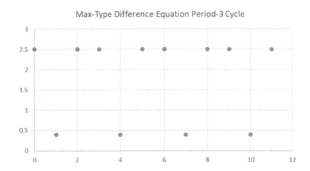

FIGURE 5.3: Descending Triangular-Shaped Period-3 Cycle.

The diagram below (Figure 5.4) depicts an eventually periodic solution with period-3 with two transient terms when $C = 1$, $x_0 = 0.8$ and $x_1 = 1.6$:

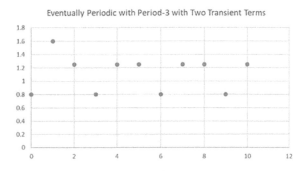

FIGURE 5.4: Eventually Periodic with Period-3 with Two Transient Terms.

The periodic essence of Max-Type Difference Equations was investigated by several authors in [1,4,5,8,10]. Now we will assume that $C < 1$ and show that every solution of Eq. (5.3) is either periodic with period-2, eventually periodic with period-2 or eventually constant.

5.1.1   Eventually Periodic with Period-2

Assume that $C < 1$. First we will show that Eq. (5.3) has no period-4 cycles.

Theorem 5.1 *Suppose that $C < 1$. Then Eq. (5.3) has no period-4 cycles.*

Proof: *We will show that there exists $N \geq 0$ such that*

$$x_{N+1} = \frac{1}{x_N}.$$

For the sake of contradiction, suppose that $x_{n+2} = \frac{C}{x_n}$ for all $n \geq 0$. Then by iterations we procure the cognate properties:

$$x_0,$$

$$x_1,$$

$$x_2 = max\left\{\frac{1}{x_1}, \frac{C}{x_0}\right\} = \frac{C}{x_0} \quad \left(\text{if } \frac{x_0}{x_1} < C < 1\right),$$

$$x_3 = max\left\{\frac{1}{[x_2]}, \frac{C}{x_1}\right\} = max\left\{\frac{1}{\left[\frac{C}{x_0}\right]}, \frac{C}{x_1}\right\} = max\left\{\frac{x_0}{C}, \frac{C}{x_1}\right\} \qquad (5.4)$$

$$= \frac{C}{x_1} \quad \left(\text{if } x_0 x_1 < C^2 < 1\right),$$

$$x_4 = max\left\{\frac{1}{[x_3]}, \frac{C}{[x_2]}\right\} = max\left\{\frac{1}{\left[\frac{C}{x_1}\right]}, \frac{C}{\left[\frac{C}{x_0}\right]}\right\} = max\left\{\frac{x_1}{C}, x_0\right\}$$

$$= x_0 \quad \left(\text{if } \frac{x_1}{x_0} < C < 1\right).$$

Via (5.4), we obtain the corresponding inequalities:

$$\frac{x_0}{x_1} < C \quad \text{and} \quad \frac{x_1}{x_0} < C,$$

which is clearly a contradiction. Thus Eq. (5.3) cannot have period-4 cycles when $C < 1$. Therefore, when $C < 1$, it suffices to consider the following initial conditions:

$$x_0 \quad \text{and} \quad x_1 = \frac{1}{x_0}. \qquad (5.5)$$

The consequent example will outline the necessary and sufficient conditions for every solution of Eq. (5.3) to be periodic with period-2.

Example 5.1 *Suppose that $C < 1$. Then Eq. (5.3) has period-2 cycles if and only if*

$$C < x_0^2 < \frac{1}{C}. \tag{5.6}$$

Solution: *Via Theorem 5.1 and (5.5), let $x_1 = \frac{1}{x_0}$. Then we obtain:*

$$x_0,$$

$$x_1 = \frac{1}{x_0},$$

$$x_2 = max\left\{\frac{1}{[x_1]}, \frac{C}{x_0}\right\} = max\left\{\frac{1}{\left[\frac{1}{x_0}\right]}, \frac{C}{x_0}\right\} = max\left\{x_0, \frac{C}{x_0}\right\}$$

$$= x_0 \quad (\text{if } x_0^2 > C),$$

$$x_3 = max\left\{\frac{1}{[x_2]}, \frac{C}{[x_1]}\right\} = max\left\{\frac{1}{x_0}, \frac{C}{\left[\frac{1}{x_0}\right]}\right\} = max\left\{\frac{1}{x_0}, Cx_0\right\} \tag{5.7}$$

$$= \frac{1}{x_0} \quad \left(\text{if } x_0^2 < \frac{1}{C}\right).$$

Via (5.7), we obtain period-2 cycles if and only if (5.6) holds.

Therefore, via (5.6) either if $x_0^2 < C$ or if $x_0^2 > \frac{1}{C}$, then Eq. (5.3) will have eventually periodic cycles with period-2. Then the fundamental question to ask: exactly how many transient terms will Eq. (5.3) have and under what criteria? The next three examples will remit these specific details and the pattern of the transient terms when $x_0^2 < C$. The case when $x_0^2 > \frac{1}{C}$ will be remitted later in this section.

Example 5.2 *Suppose that $C < 1$ and*

$$C^3 < x_0^2 < C. \tag{5.8}$$

Then Eq. (5.3) is eventually periodic with period-2 with two transient terms.

Solution: *Via Theorem 5.1 and (5.5) let $x_1 = \frac{1}{x_0}$. Then we acquire:*

$$x_0,$$

$$x_1 = \frac{1}{x_0},$$

$$x_2 = max\left\{\frac{1}{[x_1]}, \frac{C}{x_0}\right\} = max\left\{\frac{1}{\left[\frac{1}{x_0}\right]}, \frac{C}{x_0}\right\} = max\left\{x_0, \frac{C}{x_0}\right\}$$

$$= \frac{\mathbf{C}}{\mathbf{x_0}} \quad (\text{as } x_0^2 < C),$$

$$x_3 = max\left\{\frac{1}{[x_2]}, \frac{C}{[x_1]}\right\} = max\left\{\frac{1}{\left[\frac{C}{x_0}\right]}, \frac{C}{\left[\frac{1}{x_0}\right]}\right\} = max\left\{\frac{x_0}{C}, Cx_0\right\}$$

$$= \frac{\mathbf{x_0}}{\mathbf{C}} \quad (as \ \ C < 1),$$

$$x_4 = max\left\{\frac{1}{[x_3]}, \frac{C}{[x_2]}\right\} = max\left\{\frac{1}{\left[\frac{x_0}{C}\right]}, \frac{C}{\left[\frac{C}{x_0}\right]}\right\} = max\left\{\frac{C}{x_0}, x_0\right\}$$

$$= \frac{\mathbf{C}}{\mathbf{x_0}} = x_2 \quad (as \ \ x_0^2 < C),$$

$$x_5 = max\left\{\frac{1}{[x_4]}, \frac{C}{[x_3]}\right\} = max\left\{\frac{1}{\left[\frac{C}{x_0}\right]}, \frac{C}{\left[\frac{x_0}{C}\right]}\right\} = max\left\{\frac{x_0}{C}, \frac{C^2}{x_0}\right\}$$

$$= \frac{\mathbf{x_0}}{\mathbf{C}} = x_3 \quad (if \ \ C^3 < x_0^2 < C). \tag{5.9}$$

Via (5.8) and (5.9), we obtain the two transient terms in square brackets:

$$\left[x_0, \frac{1}{x_0}\right], \frac{C}{x_0}, \frac{x_0}{C}, \dots \tag{5.10}$$

Also via (5.10), for all $n \geq 0$ we acquire the corresponding period-2 pattern after pattern at the end of the sentence

$$x_{2+2n} = \frac{C}{x_0} \quad and \quad x_{3+2n} = \frac{x_0}{C}. \tag{5.11}$$

Example 5.3 *Suppose that $C < 1$ and*

$$C^5 < x_0^2 < C^3. \tag{5.12}$$

Then Eq. (5.3) is eventually periodic with period-2 with five transient terms.

Solution: *Via Theorem 5.1 and (5.5) let $x_1 = \frac{1}{x_0}$. Then we obtain:*

$$x_0,$$

$$x_1 = \frac{1}{x_0},$$

$$x_2 = max\left\{\frac{1}{[x_1]}, \frac{C}{x_0}\right\} = max\left\{x_0, \frac{C}{x_0}\right\} = \frac{C}{x_0},$$

$$(as \ \ x_0^2 < C),$$

$$x_3 = max\left\{\frac{1}{[x_2]}, \frac{C}{[x_1]}\right\} = max\left\{\frac{x_0}{C}, Cx_0\right\} = \frac{x_0}{C},$$

$$(as \ \ C < 1),$$

$$x_4 = max\left\{\frac{1}{[x_3]}, \frac{C}{[x_2]}\right\} = max\left\{\frac{C}{x_0}, x_0\right\} = \frac{C}{x_0}, \tag{5.13}$$
$$\left(\text{as } x_0^2 < C\right),$$

$$x_5 = max\left\{\frac{1}{[x_4]}, \frac{C}{[x_3]}\right\} = max\left\{\frac{x_0}{C}, \frac{C^2}{x_0}\right\} = \frac{\mathbf{C^2}}{\mathbf{x_0}},$$
$$\left(\text{as } x_0^2 < C^3\right),$$

$$x_6 = max\left\{\frac{1}{[x_5]}, \frac{C}{[x_4]}\right\} = max\left\{\frac{x_0}{C^2}, x_0\right\} = \frac{\mathbf{x_0}}{\mathbf{C^2}},$$
$$\left(\text{as } C < 1\right),$$

$$x_7 = max\left\{\frac{1}{[x_6]}, \frac{C}{[x_5]}\right\} = max\left\{\frac{C^2}{x_0}, \frac{x_0}{C}\right\} = \frac{\mathbf{C^2}}{\mathbf{x_0}} = x_5,$$
$$\left(\text{as } x_0^2 < C^3\right),$$

$$x_8 = max\left\{\frac{1}{[x_7]}, \frac{C}{[x_6]}\right\} = max\left\{\frac{x_0}{C^2}, \frac{C^3}{x_0}\right\} = \frac{\mathbf{x_0}}{\mathbf{C^2}} = x_6.$$
$$\left(\text{if } C^5 < x_0^2 < C^3\right).$$

Via (5.12) and (5.13), we obtain the five transient terms in square brackets:

$$\left[x_0, \frac{1}{x_0}, \frac{C}{x_0}, \frac{x_0}{C}, \frac{C}{x_0}\right], \frac{C^2}{x_0}, \frac{x_0}{C^2}, \dots \tag{5.14}$$

The corresponding diagram (Figure 5.5) depicts (5.12) and (5.14) when $C = 0.5$, $x_0 = \frac{1}{5}$ *and* $x_1 = 2$:

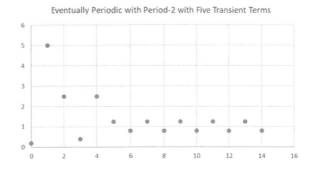

FIGURE 5.5: Eventually Periodic with Period-2 with Five Transient Terms.

We can obtain an eventually constant solution of Eq. (5.3) when either $x_0 = C^3$ *or* $x_0 = C^5$. *The sketch below (Figure 5.6) renders an eventually constant solution with eight transient terms when* $C = 0.5$, $x_0 = \frac{1}{8} = C^3$ *and* $x_1 = 8 = \frac{1}{C^3}$:

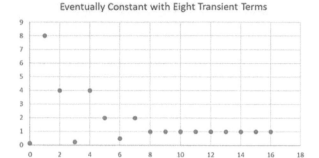

FIGURE 5.6: Eventually Constant with Eight Transient Terms.

Example 5.4 *Suppose that $C < 1$ and*

$$C^7 < x_0^2 < C^5. \tag{5.15}$$

Then Eq. (5.3) is eventually periodic with period-2 with eight transient terms.

Solution: *Analogous to (5.9), (5.13), (5.10) and (5.14), by iterations, we obtain the corresponding pattern of the eight transient terms ($x_0 - x_7$) in square brackets:*

$$\left[x_0, \frac{1}{x_0}, \frac{C}{x_0}, \frac{x_0}{C}, \frac{C}{x_0}, \frac{C^2}{x_0}, \frac{x_0}{C^2}, \frac{C^2}{x_0} \right], \tag{5.16}$$

and for all $n \geq 0$ the following period-2 pattern:

$$x_{8+2n} = \frac{C^3}{x_0} \quad and \quad x_{9+2n} = \frac{x_0}{C^3}. \tag{5.17}$$

Notice that the period-2 pattern in (5.17) is expressed as reciprocals of x_0 and C^3. Analogous pattern of the period-2 cycle in (5.9) as reciprocals of x_0 and C and of the period-2 cycle in (5.13) as reciprocals of x_0 and C^2. Alternatively we can reformulate (5.16) as the following piece-wise sequence assembled with three geometric sub-sequences:

$$x_i = \begin{cases} \frac{x_0}{C^{\frac{i}{3}}} & \text{if } i \in [0, 3, 6], \\[2mm] \frac{C^{\frac{i-1}{3}}}{x_0} & \text{if } i \in [1, 4, 7], \\[2mm] \frac{C^{\frac{i+1}{3}}}{x_0} & \text{if } i \in [2, 5]. \end{cases} \tag{5.18}$$

As $C < 1$, then the blue sub-sequence in (5.18) is an increasing geometric sub-sequence while the green and the red sub-sequences in (5.18) are decreasing geometric sub-sequences. The cognate sketch (Figure 5.7) renders (5.15), (5.18), and (5.17) when $C = 0.5$, $x_0 = \frac{1}{7}$ and $x_1 = 7$.

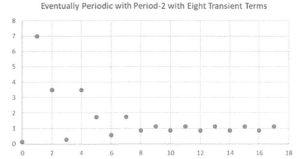

FIGURE 5.7: Eventually Periodic with Period-2 with Eight Transient Terms.

Parallel to Figure 5.6 in Example (5.3), we obtain an eventually constant solution of Eq. (5.3) when either $x_0 = C^5$ or $x_0 = C^7$. The consequent diagram (Figure 5.8) depicts an eventually constant solution with fourteen transient terms when $C = 0.5$, $x_0 = \frac{1}{32} = C^5$ and $x_1 = 32 = \frac{1}{C^5}$:

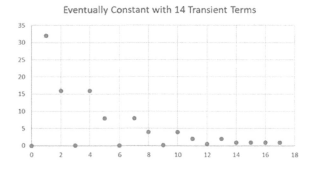

FIGURE 5.8: Eventually Constant with Fourteen Transient Terms.

From (5.8), (5.12) and (5.15), there exists $k \in \mathbb{N}$ such that:

$$C^{2k+1} < x_0^2 < C^{2k-1}, \tag{5.19}$$

resembled with the corresponding diagram:

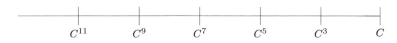

FIGURE 5.9: We choose $x_0^2 \in \left[C^{2k+1},\ C^{2k-1}\right]$ for some $k \in \mathbb{N}$.

If $x_0 = C^k$ and $x_1 = \frac{1}{C^k}$ for some $k \in \mathbb{N}$, then Eq. (5.3) will be eventually constant with $3k - 1$ transient terms as we encountered in Figures 5.6 and 5.8. This will be proved in the later section of this chapter.

Now Examples (5.2), (5.3) and (5.4) guide us to the corresponding theorem.

Theorem 5.2 *Suppose that $C < 1$, (5.19) for some $k \in \mathbb{N}$ and $x_0 \neq C^k$ for all $k \geq 2$. Then Eq. (5.3) is eventually periodic with period-2 with the corresponding $3k - 1$ transient terms:*

$$
x_i = \begin{cases} \dfrac{x_0}{C^{\frac{i}{3}}} & \text{if } i \in [0, 3, 6, \dots, 3(k-1)], \\[2ex] \dfrac{C^{\frac{i-1}{3}}}{x_0} & \text{if } i \in [1, 4, 7, \dots, 3(k-1)+1], \\[2ex] \dfrac{C^{\frac{i+1}{3}}}{x_0} & \text{if } i \in [2, 5, \dots, 3(k-1)-1], \end{cases} \tag{5.20}
$$

and for all $n \geq 0$ the cognate period-2 pattern:

$$
x_{3k-1+2n} = \frac{C^k}{x_0} \quad \text{and} \quad x_{3k+2n} = \frac{x_0}{C^k}. \tag{5.21}
$$

Notice that the pattern of (5.20) emerges from (5.18), (5.14) and (5.10), while the pattern of (5.21) is a consequence of (5.17), (5.13) and (5.5). This will be left as an end-of-chapter exercise to prove.

Now suppose that $x_0^2 > \frac{1}{C}$. Then parallel to (5.8), (5.12) and (5.15) in Examples (5.2), (5.3) and (5.4), we will come upon comparable intervals and analogous patterns of transient terms and period-2 cycles. The upcoming examples will render these contrasts such as the number of transient terms and their related patterns.

Example 5.5 *Suppose that $C < 1$ and*

$$
\frac{1}{C} < x_0^2 < \frac{1}{C^3}. \tag{5.22}
$$

Then Eq. (5.3) is eventually periodic with period-2 with three transient terms.

Solution: *Similar to Examples (5.2), (5.3) and (5.4), by iterations we obtain the cognate pattern of the three transient terms $(x_0 - x_2)$ in square brackets:*

$$
\left[x_0, \frac{1}{x_0}, x_0 \right], \ Cx_0, \ \frac{1}{Cx_0}, \ \dots, \tag{5.23}
$$

and for all $n \geq 0$ the related period-2 pattern:

$$
x_{3+2n} = Cx_0 \quad \text{and} \quad x_{4+2n} = \frac{1}{Cx_0}. \tag{5.24}
$$

The cognate sketch (Figure 5.10) describes (5.23) and (5.24) when $C = 0.5$, $x_0 = 2.5$ and $x_1 = 0.4$:

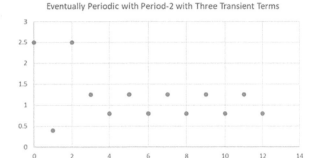

FIGURE 5.10: Eventually Periodic with Period-2 with Three Transient Terms.

Observe that the three transient terms in Figure 5.10 emerge in **descending triangular** *shape.*

Example 5.6 *Suppose that $C < 1$ and*

$$\frac{1}{C^3} < x_0^2 < \frac{1}{C^5}. \tag{5.25}$$

Then Eq. (5.3) is eventually periodic with period-2 with six transient terms.

Solution: *Similar to Example (5.5), by iterations we procure the following pattern of the six transient terms $(x_0 - x_5)$ in square brackets:*

$$\left[x_0, \frac{1}{x_0}, x_0, Cx_0, \frac{1}{Cx_0}, Cx_0 \right], \; C^2 x_0, \frac{1}{C^2 x_0}, \ldots, \tag{5.26}$$

and for all $n \geq 0$ the cognate period-2 pattern:

$$x_{6+2n} = C^2 x_0 \quad \text{and} \quad x_{7+2n} = \frac{1}{C^2 x_0}. \tag{5.27}$$

The period-2 pattern in (5.27) is expressed as $C^2 x_0$ and its reciprocal $\frac{1}{C^2 x_0}$. Parallel to (5.18) in Example (5.4), we can also recast (5.27) as a piece-wise sequence with three geometric sub-sequences:

$$x_i = \begin{cases} C^{\frac{i}{3}} x_0 & \text{if } i \in [0,3], \\ \frac{1}{C^{\frac{i-1}{3}} x_0} & \text{if } i \in [1,4], \\ C^{\frac{i-2}{3}} x_0 & \text{if } i \in [2,5]. \end{cases} \tag{5.28}$$

In comparison to (5.18) in Example (5.4), since $C < 1$ then the blue and the red sub-sequences in (5.28) are decreasing geometric sub-sequences while the green sub-sequence in (5.28) is an increasing geometric sub-sequence. The consequent diagram (Figure 5.11) traces (5.26), (5.27) and (5.28) when $C = 0.5$, $x_0 = 5$ and $x_1 = 0.2$:

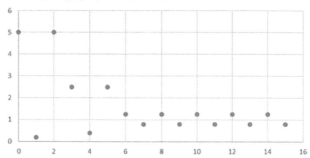

FIGURE 5.11: Eventually Periodic with Period-2 with Six Transient Terms.

The transient terms in Figure 5.11 also emerge in **descending triangular** *shapes. Parallel to Figures 5.6 and 5.8, we procure an eventually constant solution of Eq. (5.3) when either* $x_0 = \frac{1}{C^3}$ *or* $x_0 = \frac{1}{C^3}$. *The graph below (Figure 5.12) depicts an eventually constant solution with nine transient terms as* **descending triangular shapes** *when* $C = 0.5$, $x_0 = 8 = C^3$ *and* $x_1 = \frac{1}{8} = \frac{1}{C^3}$:

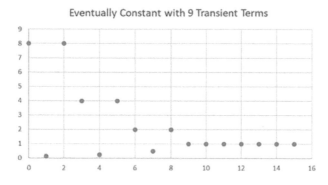

FIGURE 5.12: Eventually Constant with Nine Transient Terms.

Via (5.22) and (5.25) and analogous to (5.19), there exists $k \in \mathbb{N}$ such that:

$$\frac{1}{C^{2k-1}} < x_0^2 < \frac{1}{C^{2k+1}}. \tag{5.29}$$

resembled with the corresponding diagram (Figure 5.13):

FIGURE 5.13: We choose $x_0^2 \in \left[\frac{1}{C^{2k-1}}, \frac{1}{C^{2k+1}} \right]$ for some $k \in \mathbb{N}$.

If $x_0 = \frac{1}{C^k}$ and $x_1 = C^k$ for some $k \in \mathbb{N}$ then Eq. (5.3) will be eventually constant with $3k$ transient terms as we encountered in Figures 5.12 and 5.13. This will be proved in the later section of this chapter.

The associated theorem is a direct consequence of Examples (5.5) and (5.6).

Theorem 5.3 *Suppose that $C < 1$ and (5.29) for some $k \in \mathbb{N}$ and $x_0 \neq \frac{1}{C^k}$ for all $k \in \mathbb{N}$. Then Eq. (5.3) is eventually periodic with period-2 with the corresponding $3k$ transient terms:*

$$
x_i = \begin{cases} C^{\frac{i}{3}} x_0 & \text{if } i \in [0, 3, 6, \ldots, 3(k-1)], \\ \dfrac{1}{C^{\frac{i-1}{3}} x_0} & \text{if } i \in [1, 4, 7, \ldots, 3(k-1)+1], \\ C^{\frac{i-2}{3}} x_0 & \text{if } i \in [2, 5, \ldots, 3(k-1)+2], \end{cases} \tag{5.30}
$$

and for all $n \geq 0$ the associated period-2 pattern:

$$
x_{3k+2n} = C^k x_0 \quad \text{and} \quad x_{3k+1+2n} = \frac{1}{C^k x_0}. \tag{5.31}
$$

The pattern of (5.30) follows from (5.23), (5.26) and (5.28), while the pattern of (5.31) follows from (5.24), (5.27) and (5.29). This will be left as an end-of-chapter exercise to prove.

5.1.2 Eventually Periodic with Period-4

Now we will assume that $C > 1$ and show that every solution of Eq. (5.3) is either periodic with period-4 or eventually periodic with period-4. First we will show that Eq. (5.3) has no period-2 cycles.

Theorem 5.4 *Suppose that $C > 1$. Then Eq. (5.3) has no period-2 cycles.*

Proof: *We will show that there exists $N \geq 0$ such that*

$$
x_{N+2} = \frac{C}{x_N}.
$$

For the sake of contradiction, suppose that $x_{n+1} = \frac{1}{x_n}$ for all $n \geq 0$. Then by iterations we acquire the following properties:

$x_0,$

$x_1 = \dfrac{1}{x_0},$

$x_2 = max\left\{\dfrac{1}{x_1}, \dfrac{C}{x_0}\right\} = max\left\{x_0, \dfrac{C}{x_0}\right\} = x_0 \quad (\text{if } x_0^2 \geq C > 1),$ (5.32)

$x_3 = max\left\{\dfrac{1}{x_2}, \dfrac{C}{x_1}\right\} = max\left\{\dfrac{1}{x_0}, C x_0\right\} = \dfrac{1}{x_0} \quad (\text{if } C x_0^2 \leq 1).$

Via (5.32), we procure the corresponding inequality:

$$1 < C \le x_0^2 \le \frac{1}{C} < 1,$$

which is clearly a contradiction as we assumed that $C > 1$. Thus Eq. (5.3) cannot have period-2 cycles when $C > 1$.

The upcoming example will indicate the necessary and sufficient conditions for every solution of Eq. (5.3) to be periodic with period-4.

Example 5.7 *Suppose that $C > 1$. Then Eq. (5.3) has period-4 cycles if and only if*

$$\frac{1}{C} < x_0^2 < C. \tag{5.33}$$

Solution: *Via Theorem 5.4 and (5.5), let $x_1 = \frac{1}{x_0}$. Then we obtain:*

$$x_0,$$

$$x_1 = \frac{1}{x_0},$$

$$x_2 = max\left\{\frac{1}{x_1}, \frac{C}{x_0}\right\} = max\left\{x_0, \frac{C}{x_0}\right\} = \frac{C}{x_0} \quad (\text{if } x_0^2 < C),$$

$$x_3 = max\left\{\frac{1}{x_2}, \frac{C}{x_1}\right\} = max\left\{\frac{x_0}{C}, Cx_0\right\} = Cx_0 \quad (\text{as } C > 1),$$

$$x_4 = max\left\{\frac{1}{x_3}, \frac{C}{x_2}\right\} = max\left\{\frac{1}{Cx_0}, x_0\right\} = x_0 \quad \left(\text{if } x_0^2 > \frac{1}{C}\right),$$

$$x_5 = max\left\{\frac{1}{x_4}, \frac{C}{x_3}\right\} = max\left\{\frac{1}{x_0}, \frac{1}{x_0}\right\} = \frac{1}{x_0} = x_1. \tag{5.34}$$

Via (5.34), we obtain period-4 cycles if and only if (5.33) holds.

Hence, via (5.33) if either $x_0^2 < \frac{1}{C}$ or $x_0^2 > C$, then Eq. (5.3) will have eventually periodic solutions with period-4. Parallel to the previous section, if $x_0^2 < \frac{1}{C}$ then Eq. (5.3) will be eventually periodic with period-4 with $3k$ transient terms ($k \in \mathbb{N}$). On the contrary, if $x_0^2 > C$ then Eq. (5.3) will be eventually periodic with period-4 with $3k + 1$ transient terms ($k \in \mathbb{N}$).

Furthermore, if $x_0 = C^k$ or if $x_0 = \frac{1}{C^k}$ ($k \in \mathbb{N}$) then Eq. (5.3) will not exhibit eventually constant solutions as we encountered in the previous section.

The next series examples will remit these specific details and the pattern of the transient terms when $x_0^2 < \frac{1}{C}$.

Example 5.8 *Suppose that $C > 1$ and*

$$\frac{1}{C^3} < x_0^2 < \frac{1}{C}. \tag{5.35}$$

Then Eq. (5.3) is eventually periodic with period-4 with three transient terms.

Solution: *By iterations we obtain the corresponding pattern of the three transient terms $(x_0 - x_2)$ in square brackets:*

$$\left[x_0, \frac{1}{x_0}, \frac{C}{x_0}\right], \ Cx_0, \ \frac{1}{Cx_0}, \ \frac{1}{x_0}, \ C^2x_0, \ \ldots, \tag{5.36}$$

and for all $n \geq 0$ the following period-4 pattern:

$$\begin{cases} x_{3+4n} = Cx_0, \\ x_{4+4n} = \frac{1}{Cx_0}, \\ x_{5+4n} = \frac{1}{x_0}, \\ x_{6+4n} = C^2x_0. \end{cases} \tag{5.37}$$

The cognate sketch (Figure 5.14) depicts (5.36) and (5.37) when $C = 2$, $x_0 = 0.4$ and $x_1 = 2.5$:

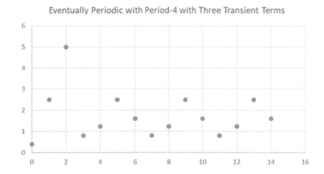

Eventually Periodic with Period-4 with Three Transient Terms

FIGURE 5.14: Eventually Periodic with Period-4 with Three Transient Terms.

Figure 5.14 renders a sequence of three ascending transient terms.

Example 5.9 *Suppose that $C > 1$ and*

$$\frac{1}{C^5} < x_0^2 < \frac{1}{C^3}. \tag{5.38}$$

Then Eq. (5.3) is eventually periodic with period-4 with six transient terms.

Solution: *We acquire the corresponding pattern of the six transient terms $(x_0 - x_5)$ in square brackets:*

$$\left[x_0, \frac{1}{x_0}, \frac{C}{x_0}, Cx_0, \frac{1}{Cx_0}, \frac{1}{x_0}\right], \tag{5.39}$$

and for all $n \geq 0$ the cognate period-4 pattern:

$$\begin{cases} x_{6+4n} = C^2 x_0, \\ x_{7+4n} = \frac{1}{C^2 x_0}, \\ x_{8+4n} = \frac{1}{C x_0}, \\ x_{9+4n} = C^3 x_0. \end{cases} \tag{5.40}$$

The graph below (Figure 5.15) evokes (5.39) and (5.40) when $C = 2$, $x_0 = 0.3$ and $x_1 = \frac{10}{3}$:

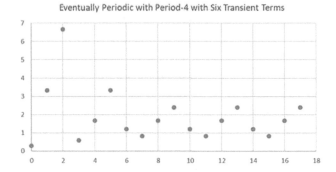

Eventually Periodic with Period-4 with Six Transient Terms

FIGURE 5.15: Eventually Periodic with Period-4 with Six Transient Terms.

Figure 5.15 renders the transient terms as a piece-wise sequence composed with three sub-sequences. Alternatively, we revise (5.39) as a piece-wise sequence with three geometric sub-sequences:

$$x_i = \begin{cases} C^{\frac{i}{3}} x_0 & \text{if } i \in [0,3], \\ \dfrac{1}{C^{\frac{i-1}{3}} x_0} & \text{if } i \in [1,4], \\ \dfrac{C^{\frac{5-i}{3}}}{x_0} & \text{if } i \in [2,5]. \end{cases} \tag{5.41}$$

In Figure 5.15 and (5.41), the blue geometric sub-sequence is an increasing sub-sequence, while the green and the red geometric sub-sequences are decreasing sub-sequences.

Analogous to Figures 5.14 and 5.15, the sketch below (Figure 5.16) renders an eventually periodic cycle with period-4 with nine transient terms when $C = 2$, $x_0 = 0.1$ and $x_1 = 10$:

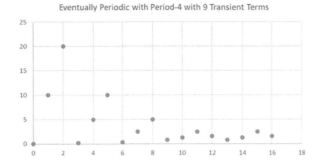

FIGURE 5.16: Eventually Periodic with Period-4 with Nine Transient Terms.

Via (5.35) and (5.38) and analogous to (5.29), there exists $k \in \mathbb{N}$ such that:

$$\frac{1}{C^{2k+1}} < x_0^2 < \frac{1}{C^{2k-1}}.\qquad(5.42)$$

portrayed with the associated diagram (Figure 5.17):

FIGURE 5.17: We choose $x_0^2 \in \left[\frac{1}{C^{2k+1}}, \frac{1}{C^{2k-1}}\right]$ for some $k \in \mathbb{N}$.

The corresponding theorem follows directly from Examples (5.8) and (5.9).

Theorem 5.5 *Suppose that $C > 1$ and (5.42) for some $k \in \mathbb{N}$. Then Eq. (5.3) is eventually periodic with period-4 with the corresponding $3k$ transient terms:*

$$x_i = \begin{cases} C^{\frac{i}{3}}x_0 & \text{if } i \in [0,3,6,\ldots,3(k-1)], \\ \dfrac{1}{C^{\frac{i-1}{3}}x_0} & \text{if } i \in [1,4,7,\ldots,3(k-1)+1], \\ C^{\frac{5-i}{3}}x_0 & \text{if } i \in [2,5,\ldots,3(k-1)+2], \end{cases}\qquad(5.43)$$

and for all $n \geq 0$ the cognate period-4 pattern:

$$\begin{cases} x_{3k+4n} = C^k x_0, \\ x_{3k+1+4n} = \dfrac{1}{C^k x_0}, \\ x_{3k+2+4n} = \dfrac{1}{C^{k-1}x_0}, \\ x_{3k+3+4n} = C^{k+1}x_0. \end{cases}\qquad(5.44)$$

The pattern of (5.43) transitions from (5.36), (5.39) and (5.41), while the pattern of (5.44) follows from (5.37), (5.40) and (5.42). This will be left as an end-of-chapter exercise to prove.

Now suppose that $x_0^2 > C$. Then analogous to (5.35) and (5.38) in Examples (5.8) and (5.9) we will determine comparable intervals and analogous patterns of the transient terms and period-4 cycles. The succeeding examples will portray these contrasts, especially will show that Eq. (5.3) will be eventually periodic with period-4 with $3k + 1$ transient terms ($k \in \mathbb{N}$), in comparison to $3k$ transient terms when $x_0^2 < \frac{1}{C}$.

Example 5.10 *Suppose that $C > 1$ and*

$$C < x_0^2 < C^3. \tag{5.45}$$

Then Eq. (5.3) is eventually periodic with period-4 with four transient terms.

Solution: *We procure the cognate pattern of the four transient terms $(x_0 - x_3)$ in square brackets:*

$$\left[x_0, \frac{1}{x_0}, x_0, Cx_0 \right], \frac{C}{x_0}, \frac{x_0}{C}, x_0, \frac{C^2}{x_0}, \ldots, \tag{5.46}$$

and for all $n \geq 0$ the following period-4 pattern:

$$\begin{cases} x_{4+4n} = \frac{C}{x_0}, \\ x_{5+4n} = \frac{x_0}{C}, \\ x_{6+4n} = x_0, \\ x_{7+4n} = \frac{C^2}{x_0}. \end{cases} \tag{5.47}$$

Depending on the values of C, x_0 and x_1, (5.47) can emerge in various shapes. On the other hand, the pattern of the transient terms in (5.46) will be similar to what we observed in the previous section (two decreasing sub-sequences and one increasing sub-sequence). In fact, in (5.46) x_0 is the only isolated term and $x_1 - x_3$ emerge as a piece-wise sequence with three sub-sequences.

*The cognate sketch (Figure 5.18) depicts (5.46) and (5.47) with four transient terms and an **ascending step-shaped** period-4 cycle when $C = 2$, $x_0 = 2$ and $x_1 = 0.5$:*

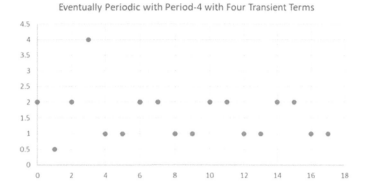

FIGURE 5.18: Eventually Periodic with Period-4 with Four Transient Terms.

Example 5.11 *Suppose that $C > 1$ and*

$$C^3 < x_0^2 < C^5. \tag{5.48}$$

Then Eq. (5.3) is eventually periodic with period-4 with seven transient terms.

Solution: *We obtain the corresponding pattern of the seven transient terms $(x_0 - x_6)$ in square brackets:*

$$\left[x_0, \frac{1}{x_0}, x_0, Cx_0, \frac{C}{x_0}, \frac{x_0}{C}, x_0 \right], \frac{C^2}{x_0}, \frac{x_0}{C^2}, \frac{x_0}{C}, \frac{C^3}{x_0}, \dots, \tag{5.49}$$

and for all $n \geq 0$ the associated period-4 pattern:

$$\begin{cases} x_{7+4n} = \frac{C^2}{x_0}, \\ x_{8+4n} = \frac{x_0}{C^2}, \\ x_{9+4n} = \frac{x_0}{C}, \\ x_{10+4n} = \frac{C^3}{x_0}. \end{cases} \tag{5.50}$$

*The next graph (Figure 5.19) describes (5.49) and (5.50) with seven transient terms and an **ascending step-shaped** period-4 cycle when $C = 2$, $x_0 = 4$ and $x_1 = 0.25$:*

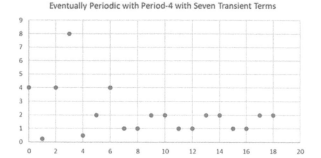

FIGURE 5.19: Eventually Periodic with Period-4 with Seven Transient Terms.

Observe that in Figure 5.19 the first term x_0 is an isolated term and the rest of the terms $(x_1 - x_6)$ emerge in three geometric sub-sequences as portrayed in (5.49). Now we reformulate (5.49) as a piece-wise sequence with three geometric sub-sequences with the first isolated term x_0:

$$x_i = \begin{cases} x_0 & \text{if } i = 0, \\ \dfrac{C^{\frac{i-1}{3}}}{x_0} & \text{if } i \in [1,4], \\ \dfrac{x_0}{C^{\frac{i-2}{3}}} & \text{if } i \in [2,5], \\ C^{\frac{6-i}{3}} x_0 & \text{if } i \in [3,6]. \end{cases} \tag{5.51}$$

Note that in (5.51) the blue geometric sub-sequence is an increasing sub-sequence and the green and the red geometric sub-sequences are decreasing sub-sequences.

Via (5.45) and (5.48) and analogous to (5.19), there exists $k \in \mathbb{N}$ such that:

$$C^{2k-1} < x_0^2 < C^{2k+1} \tag{5.52}$$

rendered with the associated diagram (Figure 5.20):

FIGURE 5.20: We choose $x_0^2 \in \left[C^{2k-1}, C^{2k+1}\right]$ for some $k \in \mathbb{N}$.

The upcoming theorem results from Examples (5.10) and (5.11).

Theorem 5.6 *Suppose that $C > 1$ and (5.52) for some $k \in \mathbb{N}$. Then Eq. (5.3) is eventually periodic with period-4 with the associated $3k + 1$ transient terms:*

$$x_i = \begin{cases} x_0 & \text{if } i = 0, \\ \dfrac{C^{\frac{i-1}{3}}}{x_0} & \text{if } i \in [1, 4, \ldots, 3k-2], \\ \dfrac{x_0}{C^{\frac{i-2}{3}}} & \text{if } i \in [2, 5, \ldots, 3k-1], \\ C^{\frac{6-i}{3}} x_0 & \text{if } i \in [3, 6, \ldots, 3k]. \end{cases} \tag{5.53}$$

and for all $n \geq 0$ the corresponding period-4 pattern:

$$\begin{cases} x_{3k+1+4n} = \dfrac{C^k}{x_0}, \\ x_{3k+2+4n} = \dfrac{x_0}{C^k}, \\ x_{3k+3+4n} = \dfrac{x_0}{C^{k-1}}, \\ x_{3k+4+4n} = \dfrac{C^{k+1}}{x_0}. \end{cases} \tag{5.54}$$

The pattern of (5.53) extends from (5.46), (5.49) and (5.51), while the pattern of (5.54) follows from (5.47), (5.50) and (5.52). This will be left as an end-of-chapter exercise to prove.

5.1.3 Eventually Periodic with Period-3

Now we will assume that $C = 1$ and show that every solution of Eq. (5.3) is either periodic with period-3 or eventually periodic with period-3 with three maximum transient terms. First we will examine assorted graphical examples. The diagram below (Figure 5.21) renders a **descending triangular-shaped period-3 cycle** of Eq. (5.3) when $C = 1$, $x_0 = 2$ and $x_1 = 0.5$:

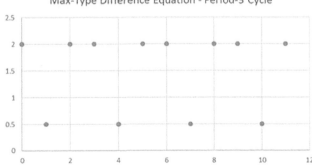

FIGURE 5.21: Descending Triangular-Shaped Period-3 Cycle.

The next sketch below (Figure 5.22) depicts an **ascending step-shaped period-3 cycle** of Eq. (5.3) when $C = 1$, $x_0 = 0.5$ and $x_1 = 2$:

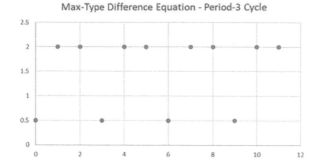

FIGURE 5.22: Ascending Step-Shaped Period-3 Cycle.

The consequent graph (Figure 5.23) portrays an eventually periodic cycle with period-3 with three ascending transient terms cycle of Eq. (5.3) when $C = 1$, $x_0 = 0.2$ and $x_1 = 2$:

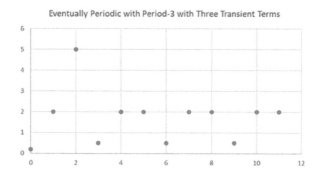

FIGURE 5.23: Eventually Periodic with Period-3 with Three Transient Terms.

Showing that when $C = 1$ that every solution of Eq. (5.3) is either periodic with period-3 or eventually periodic with period-3 will require several cases. The upcoming theorem will outline the necessary and sufficient conditions for every solution of Eq. (5.3) to be periodic with period-3.

Theorem 5.7 *Suppose that $C = 1$. Then Eq. (5.3) is periodic with period-3 if and only if one of the following holds true:*

1. $x_1 = \frac{1}{x_0} \neq 1$, or

2. $x_1 = x_0 \quad and \quad x_0^2 > 1$.

Proof: *First assume that $x_1 = \frac{1}{x_0} \neq 1$. Then by iterations we procure the associated period-3 pattern:*

$$x_0, \ \frac{1}{x_0}, \ \frac{1}{x_0}, \ \dots \tag{5.55}$$

The cognate sketch *(Figure 5.24)* renders *(5.56)* as a **descending triangular-shaped** *period-3 cycle when* $C = 1$, $x_0 = 3.2$ *and* $x_1 = 0.3125$:

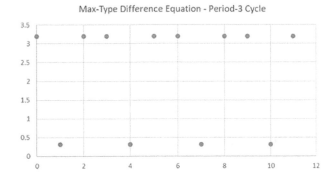

FIGURE 5.24: Descending Triangular-Shaped Period-3 Cycle.

The diagram below *(Figure 5.25)* traces *(5.56)* as an **ascending step-shaped** *period-3 cycle when* $C = 1$, $x_0 = 0.25$ *and* $x_1 = 4$:

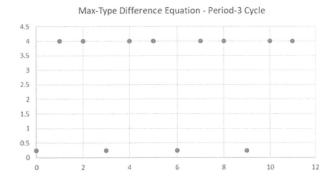

FIGURE 5.25: Ascending Step-Shaped Period-3 Cycle.

Now assume that $x_1 = x_0$ and $x_0^2 > 1$. Then we obtain the corresponding period-3 pattern:

$$x_0, \; x_0, \; \frac{1}{x_0}, \; \dots \tag{5.56}$$

The consequent graph *(Figure 5.26)* depicts *(5.56)* as a **descending step-shaped** *period-3 cycle when* $C = 1$, $x_0 = 5$ *and* $x_1 = 5$:

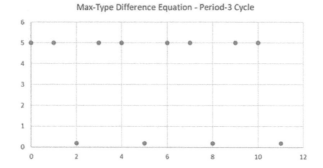

FIGURE 5.26: Descending Step-Shaped Period-3 Cycle.

On the other hand, if $x_1 = x_0$ and $x_0^2 < 1$, then Eq. (5.3) will have exactly one transient term (x_0) in square brackets depicted by the related pattern:

$$[\ \mathbf{x_0}\ ],\ x_0,\ x_0,\ \frac{1}{x_0},\ \dots \tag{5.57}$$

The sketch below (Figure 5.27) evokes (5.57) when $C = 1$, $x_0 = 0.5$ and $x_1 = 0.5$:

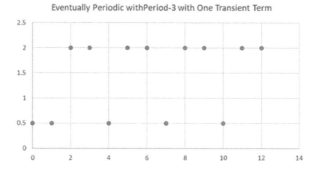

FIGURE 5.27: Eventually Periodic with Period-3 with One Transient Term.

The next theorem will outline eventually periodic solutions with period-3 of Eq. (5.3) with one transient term when $x_0 > x_1$.

Theorem 5.8 *Suppose that $C = 1$. Then Eq. (5.3) is eventually periodic with period-3 with one transient term if one of the following holds true:*

1. $x_0 > x_1$ and $x_1^2 > 1$, or

2. $x_0 > x_1$ and $x_1^2 < 1$.

Proof: *We will assume that $x_1 \neq \frac{1}{x_0}$ since every solution of Eq. (5.3) is periodic with period-3 when $x_1 = \frac{1}{x_0}$. First suppose that $x_0 > x_1$ and $x_1^2 > 1$.*

Then we obtain the corresponding pattern with one transient term (x_0):

$$[\ \mathbf{x_0} \ ], \ x_1, \ \frac{1}{x_1}, \ x_1, \ \dots \tag{5.58}$$

*The diagram below (Figure 5.28) traces (5.58) with one transient term and with a **descending triangular-shaped** period-3 cycle when $C = 1$, $x_0 = 3$ and $x_1 = 2$:*

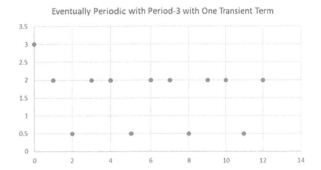

FIGURE 5.28: Eventually Periodic with Period-3 with One Transient Term.

Now assume that $x_0 > x_1$ and $x_1^2 < 1$. Then we procure the cognate pattern with one transient term (x_0):

$$[\ \mathbf{x_0} \ ], \ x_1, \ \frac{1}{x_1}, \ \frac{1}{x_1}, \ \dots \tag{5.59}$$

*The next graph (Figure 5.29) describes (5.59) with one transient term and with an **ascending step-shaped** period-3 cycle when $C = 1$, $x_0 = 3$ and $x_1 = 0.5$:*

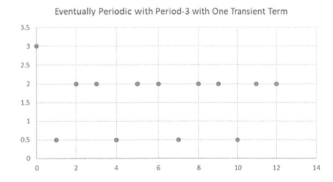

FIGURE 5.29: Eventually Periodic with Period-3 with One Transient Term.

The upcoming theorem describes eventually periodic solutions with period-3 of Eq. (5.3) with two transient terms when $x_0 < x_1$.

Theorem 5.9 *Suppose that $C = 1$. Then Eq. (5.3) is eventually periodic with period-3 with two transient terms if one of the following holds true:*

1. $x_0 < x_1$, $\quad x_0 x_1 \geq 1$ $\quad$ and $\quad x_0^2 < 1$, or

2. $x_0 < x_1$, $\quad x_0 x_1 \geq 1$ $\quad$ and $\quad x_0^2 > 1$.

Proof: *Analogous to Theorem (5.8), suppose that $x_1 \neq \frac{1}{x_0}$. First assume that $x_0 < x_1$, $x_0 x_1 \geq 1$ and $x_0^2 < 1$. Then we acquire the corresponding pattern with two transient terms in square brackets:*

$$[\, \mathbf{x_0}, \ \mathbf{x_1} \,], \ \frac{1}{x_0}, \ x_0, \ \frac{1}{x_0}, \ \ldots \tag{5.60}$$

*The sketch below (Figure 5.30) portrays (5.60) with two ascending transients terms and with a **descending triangular-shaped** period-3 cycle when $C = 1$, $x_0 = 0.4$ and $x_1 = 4$:*

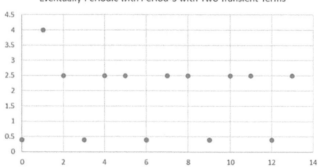

FIGURE 5.30: Eventually Periodic with Period-3 with Two Transient Terms.

Now suppose that $x_0 < x_1$, $x_0 x_1 \geq 1$ and $x_0^2 > 1$. Then we procure the cognate pattern with two transient terms in square brackets:

$$[\, \mathbf{x_0}, \ \mathbf{x_1} \,], \ \frac{1}{x_0}, \ x_0, \ x_0, \ \ldots \tag{5.61}$$

*The upcoming graph (Figure 5.31) depicts (5.61) with two ascending transient terms and with an **ascending step-shaped** period-3 cycle when $C = 1$, $x_0 = 2$ and $x_1 = 3$:*

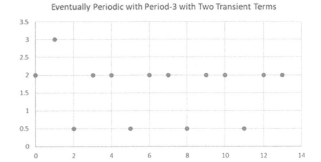

Eventually Periodic with Period-3 with Two Transient Terms

FIGURE 5.31: Eventually Periodic with Period-3 with Two Transient Terms.

The succeeding theorem portrays eventually periodic solutions with period-3 of Eq. (5.3) with three transient terms when $x_0 < x_1$.

Theorem 5.10 *Suppose that $C = 1$. Then Eq. (5.3) is eventually periodic with period-3 with three transient terms if one of the following holds true:*

1. $x_0 < x_1$, $\quad x_0 x_1 < 1$ $\quad$ and $\quad x_1^2 < 1$, or

2. $x_0 < x_1$, $\quad x_0 x_1 < 1$ $\quad$ and $\quad x_1^2 > 1$.

Proof: *First assume that $x_0 < x_1$, $x_0 x_1 < 1$ and $x_1^2 < 1$. Then we obtain the cognate pattern with three transient terms in square brackets:*

$$\left[x_0,\ x_1,\ \frac{1}{x_0} \right],\ \frac{1}{x_1},\ x_1,\ \frac{1}{x_1},\ \dots \quad (5.62)$$

The diagram below (Figure 5.32) depicts (5.62) with three ascending transients terms and with a **descending triangular-shaped** *period-3 cycle when $C = 1$, $x_0 = 0.64$ and $x_1 = 0.8$:*

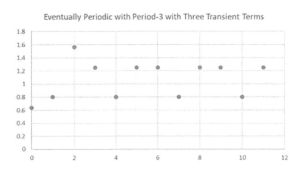

Eventually Periodic with Period-3 with Three Transient Terms

FIGURE 5.32: Eventually Periodic with Period-3 with Three Transient Terms.

Now suppose that $x_0 < x_1$, $x_0 x_1 < 1$ and $x_1^2 > 1$. Then we procure the associated pattern with three transient terms in square brackets:

$$\left[\; \mathbf{x_0}, \; \mathbf{x_1}, \; \frac{1}{\mathbf{x_0}} \; \right], \; \frac{1}{x_1}, \; x_1, \; x_1, \; \dots . \tag{5.63}$$

*The diagram below (Figure 5.33) depicts (5.63) with three ascending transients terms and with an **ascending step-shaped** period-3 cycle when $C = 1$, $x_0 = 0.4$ and $x_1 = 1.6$:*

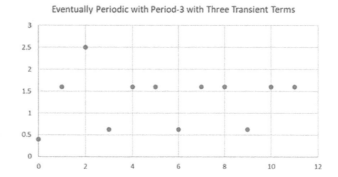

FIGURE 5.33: Eventually Periodic with Period-3 with Three Transient Terms.

Finally, it is interesting to note that if either $x_0 = 1$ or $x_0 = 1$ then Eq. (5.3) will be eventually constant with either one, two or three transient terms. The number of transient terms will vary depending on one of the following three cases:

1. $x_0 = 1$ and $x_1 > 1$,

2. $x_1 = 1$ and $x_0 < 1$,

3. $x_1 = 1$ and $x_0 > 1$.

The case when $x_0 = 1$ and $x_1 < 1$ is not included as this case will produce an eventually periodic cycle with period-3 with two descending transient terms. The upcoming graphical examples will portray cases 1 and 2. For instance, the sketch below (Figure 5.34) renders an eventually constant solution with three ascending transient terms when $C = 1$, $x_0 = 0.4$ and $x_1 = 1$:

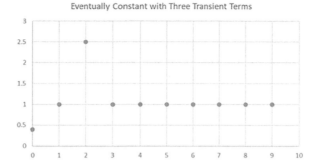

FIGURE 5.34: Eventually Constant with Three Transient Terms.

The next diagram (Figure 5.35) depicts an eventually constant solution with two ascending transient terms when $C = 1$, $x_0 = 1$ and $x_1 = 2$:

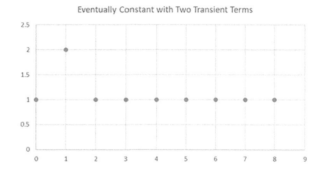

FIGURE 5.35: Eventually Constant with Two Transient Terms.

5.1.4 Eventually Constant with $K = 1$

In this section, we will assume that $C < 1$ and show that every solution of Eq. (5.3) is eventually constant with $K = 1$. First of all, recall that Figure 5.8 portrayed an eventually constant solution with $K = 1$ with fourteen transient terms when $C = 0.5$, $x_0 = \frac{1}{32} = C^5$ and $x_1 = 32 = \frac{1}{C^5}$, which then directed us to the cognate deduction that if for some $k \in \mathbb{N}$:

$$x_0 = C^k \quad \text{and} \quad x_1 = \frac{1}{C^k}, \tag{5.64}$$

then Eq. (5.3) is eventually constant with $K = 1$ with $3k - 1$ transient terms.

Second, recall that Figure 5.12 evoked an eventually constant solution with $K = 1$ with nine transient terms when $C = 0.5$, $x_0 = 8 = \frac{1}{C^3}$ and $x_1 = \frac{1}{8} = C^3$. This then guided us to the associated conclusion that if for some $k \in \mathbb{N}$:

$$x_0 = \frac{1}{C^k} \quad \text{and} \quad x_1 = C^k, \tag{5.65}$$

then Eq. (5.3) is eventually constant with $K = 1$ with 3k transient terms. The upcoming examples we will formulate and confirm these conclusions. First suppose that (5.65), which then directs us to (Figure 5.36):

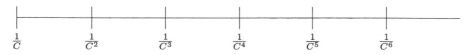

FIGURE 5.36: We choose $x_0^2 = \frac{1}{C^k}$ for some $k \in \mathbb{N}$.

Example 5.12 *Suppose that $C < 1$ and*

$$x_0 = \frac{1}{C} \quad \text{and} \quad x_1 = C. \tag{5.66}$$

Then Eq. (5.3) is eventually constant with $K = 1$ with three transient terms.

Solution: *We obtain the corresponding pattern of the three transient terms $(x_0 - x_2)$ in square brackets:*

$$\left[\frac{1}{C}, C, \frac{1}{C} \right], \ 1, \ 1, \ \ldots, \tag{5.67}$$

and for all $n \geq 0$:

$$x_{3+n} = 1. \tag{5.68}$$

*The diagram below (Figure 5.37) depicts (5.67) and (5.68) with three transient terms as a **descending triangular-shape** when $C = 0.5$, $x_0 = 2 = \frac{1}{C}$ and $x_1 = 0.5 = C$:*

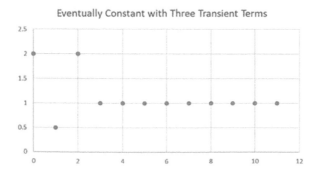

FIGURE 5.37: Eventually Constant with Three Transient Terms.

Example 5.13 *Suppose that* $C < 1$ *and*

$$x_0 = \frac{1}{C^2} \quad \text{and} \quad x_1 = C^2. \tag{5.69}$$

Then Eq. (5.3) is eventually constant with $K = 1$ *with six transient terms.*

Solution: *We procure the associated pattern of the six transient terms* $(x_0 - x_5)$ *in square brackets:*

$$\left[\frac{1}{C^2}, C^2, \frac{1}{C^2}, \frac{1}{C}, C, \frac{1}{C} \right], \ 1, \ 1, \ \ldots, \tag{5.70}$$

and for all $n \geq 0$:

$$x_{6+n} = 1. \tag{5.71}$$

Now we reformulate (5.70) as:

$$x_i = \begin{cases} \frac{1}{C^{\frac{3-i}{3}}} & \text{if } i \in [0, 3], \\ C^{\frac{1-i}{3}} & \text{if } i \in [1, 4], \\ \frac{1}{C^{\frac{5-i}{3}}} & \text{if } i \in [2, 5]. \end{cases} \tag{5.72}$$

The consequent sketch (Figure 5.38) evokes (5.70), (5.71) and (5.72) with six transient terms as **descending triangular-shapes** *when* $C = 0.5$, $x_0 = 4 = \frac{1}{C^2}$ *and* $x_1 = C^2$:

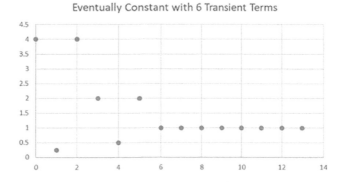

Eventually Constant with 6 Transient Terms

FIGURE 5.38: Eventually Constant with Six Transient Terms.

The upcoming sketch (Figure 5.39) portrays an eventually constant solution with twelve **descending triangular-shaped** transient terms when $C = 0.5$, $x_0 = 16 = \frac{1}{C^4}$ and $x_1 = 0.0625 = C^4$:

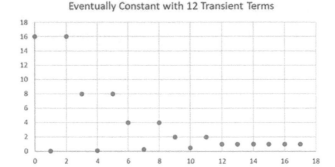

FIGURE 5.39: Eventually Constant with Twelve Transient Terms.

The succeeding theorem extends directly from Examples (5.12) and (5.13) and from Figures 5.37–5.39.

Theorem 5.11 *Suppose that $C < 1$ and (5.65) for some $k \in \mathbb{N}$. Then Eq. (5.3) is eventually constant with $K = 1$ with the associated $3k$ transient terms:*

$$
x_i = \begin{cases}
\dfrac{1}{C^{\frac{3k-3-i}{3}}} & \text{if } i \in [0, 3, \ldots, 3k-3], \\[2mm]
C^{\frac{3k-2-i}{3}} & \text{if } i \in [1, 4, \ldots, 3k-2], \\[2mm]
\dfrac{1}{C^{\frac{3k-1-i}{3}}} & \text{if } i \in [2, 5, \ldots, 3k-1].
\end{cases}
\tag{5.73}
$$

and for all $n \geq 0$:

$$
x_{3k+n} = 1.
\tag{5.74}
$$

The proof of (5.73) extends from (5.67), (5.70) and (5.72). Also (5.80) follows from (5.68) and (5.71).

Now assume that (5.64). Then analogous to Figure 5.36 we obtain (Figure 5.40):

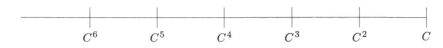

$$ C^6 \qquad C^5 \qquad C^4 \qquad C^3 \qquad C^2 \qquad C $$

FIGURE 5.40: We choose $x_0^2 = C^k$ for some $k \in \mathbb{N}$.

Example 5.14 *Suppose that $C < 1$ and*

$$
x_0 = C^2 \qquad \text{and} \qquad x_1 = \frac{1}{C^2}.
\tag{5.75}
$$

Then Eq. (5.3) is eventually constant with $K = 1$ with five transient terms.

Solution: *We acquire cognate pattern of the five transient terms $(x_0 - x_4)$ in square brackets:*

$$\left[C^2, \frac{1}{C^2}, \frac{1}{C}, C, \frac{1}{C} \right], \ 1, \ 1, \ \ldots, \tag{5.76}$$

and for all $n \geq 0$:

$$x_{5+n} = 1. \tag{5.77}$$

The consequent graph (Figure 5.41) traces (5.76) and (5.77) with five transient terms when $C = 0.5$, $x_0 = 0.25 = C^2$ and $x_1 = 4 = \frac{1}{C^2}$:

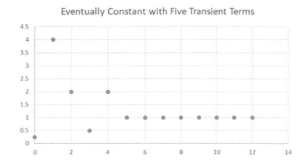

Eventually Constant with Five Transient Terms

FIGURE 5.41: Eventually Constant with Five Transient Terms.

Now we revise (5.76) as:

$$x_i = \begin{cases} \dfrac{1}{C^{\frac{3-i}{3}}} & \text{if } i \in [0,3], \\[2mm] C^{\frac{4-i}{3}} & \text{if } i \in [1,4], \\[2mm] \dfrac{1}{C^{\frac{5-i}{3}}} & \text{if } i = 2. \end{cases} \tag{5.78}$$

The next graph (Figure 5.42) renders an eventually constant solution with eleven transient terms when $C = 0.5$, $x_0 = 0.0625 = C^4$ and $x_1 = 16 = \frac{1}{C^4}$:

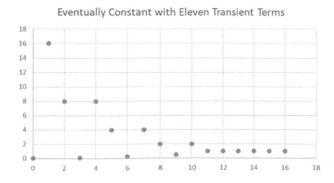

Eventually Constant with Eleven Transient Terms

FIGURE 5.42: Eventually Constant with Eleven Transient Terms.

The next theorem transitions from Example (5.14) and from Figures (5.41) and (5.42).

Theorem 5.12 *Suppose that $C < 1$ and (5.64) for some $k \geq 2$. Then Eq. (5.3) is eventually constant with $K = 1$ with the associated $3k - 1$ transient terms:*

$$x_i = \begin{cases} \dfrac{1}{C^{\frac{3k-3-i}{3}}} & \text{if } i \in [0, 3, \ldots, 3k - 3], \\ C^{\frac{3k-2-i}{3}} & \text{if } i \in [1, 4, \ldots, 3k - 2], \\ \dfrac{1}{C^{\frac{3k-4-i}{3}}} & \text{if } i \in [2, 5, \ldots, 3k - 4], \end{cases} \tag{5.79}$$

and for all $n \geq 0$:

$$x_{3k-1+n} = 1. \tag{5.80}$$

5.2   The Non-Autonomous Case (Eq. [5.2])

This section's goal is to thoroughly study the periodic traits of Eq. (5.2). We will assume that $x_0, x_1 > 0$ and $\{C_n\}_{n=0}^{\infty}$ is a period-2 sequence. Notice that the left side of Eq. (5.2) is the first-order Autonomous Riccati Difference Equation (Eq. (3.3)) and the right side of Eq. (5.2) is the second-order Non-Autonomous Riccati Difference Equation (Eq. (3.16)).

In [4], it was shown that every positive solution of Eq. (5.2) is eventually periodic with the following periods:

$$\begin{cases} 2 & \text{if } C_0 C_1 < 1, \\ 6 & \text{if } C_0 C_1 = 1, \\ 4 & \text{if } C_0 C_1 > 1. \end{cases}$$

Parallel to the previous section, we will show that the periodic cycles of Eq. (5.2) are not unique and will depend on the initial conditions x_0 and x_1 and the powers of C_0 and C_1. In addition, we will show that the transient terms of Eq. (5.2) will emerge as three piece-wise sequences with powers of C_0 and C_1. Furthermore, we will perceive intervals analogous to (5.19), (5.29), (5.42) and (5.52) as in the previous section with Eq. (5.1).

Now we will examine some graphical examples of periodic and eventually periodic solutions of Eq. (5.2) with period-2, period-4 and period-6. The upcoming sketch (Figure 5.43) renders an eventually periodic cycle with period-2 with nine **descending triangular-shaped** transient terms of Eq. (5.2) when $C_0 = 0.5$, $C_1 = 0.8$, $x_0 = 4$ and $x_1 = 0.25$:

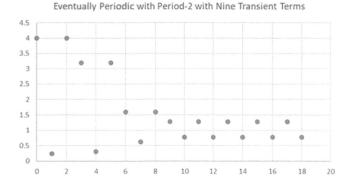

FIGURE 5.43: Eventually Periodic with Period-2 with Nine Transient Terms.

The transient terms in Figure 5.43 also emerge in **descending triangular shapes** in comparison to the transient terms in Figures 5.10 and 5.11. The next diagram (Figure 5.44) depicts an eventually periodic cycle with period-4 with seven transient terms of Eq. (5.2) when $C_0 = 3$, $C_1 = 2$, $x_0 = 8$ and $x_1 = 0.125$:

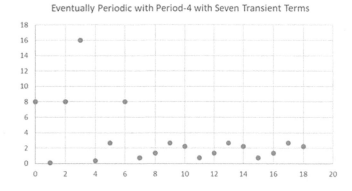

FIGURE 5.44: Eventually Periodic with Period-4 with Seven Transient Terms.

The transient terms in Figure 5.44 arise in similar patterns as the transient terms in Figures 5.18 and 5.19. The upcoming graph (Figure 5.45) describes a **descending triangular-shaped** period-6 cycle of Eq. (5.2) when $C_0 = 0.5$, $C_1 = 2$, $x_0 = 4$ and $x_1 = 0.25$:

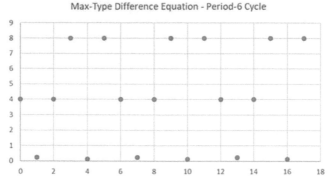

FIGURE 5.45: Descending Triangular-Shaped Period-6 Cycle.

The **descending triangular-shaped** period-6 cycle in Figure 5.45 is analogous to the **descending triangular-shaped** period-3 cycles in Figures 5.21 and 5.24. Now we will assume that $C_0C_1 < 1$ and show that every solution of Eq. (5.2) is either periodic with period-2 or eventually periodic with period-2 or eventually constant in some instances.

5.2.1 Eventually Periodic with Period-2

In this section we will assume that $C_0C_1 < 1$. Analogous to Theorem (5.1), we can show that Eq. (5.2) has no period-4 cycles. This will be left as an end-of-chapter exercise to prove. The upcoming example will outline the necessary and sufficient conditions for every solution of Eq. (5.2) to be periodic with period-2.

Example 5.15 *Suppose that $C_0C_1 < 1$. Then Eq. (5.2) has period-2 cycles if and only if*

$$C_0 < x_0^2 < \frac{1}{C_1}. \tag{5.81}$$

Solution: *Analogous to Theorem (5.1), let $x_1 = \frac{1}{x_0}$. Then we obtain:*

$$x_0,$$

$$x_1 = \frac{1}{x_0},$$

$$x_2 = max\left\{\frac{1}{[x_1]}, \frac{C_0}{x_0}\right\} = max\left\{x_0, \frac{C_0}{x_0}\right\} = x_0 \quad (\text{if } x_0^2 > C_0),$$

$$x_3 = max\left\{\frac{1}{[x_2]}, \frac{C_1}{[x_1]}\right\} = max\left\{\frac{1}{x_0}, C_1x_0\right\} = \frac{1}{x_0} = x_1.$$

$$\left(\text{if } x_0^2 < \frac{1}{C_1}\right). \tag{5.82}$$

Via (5.82), we obtain period-2 cycles if and only if (5.81) holds.

Thus, via (5.81) if either $x_0^2 > \frac{1}{C_1}$ or $x_0^2 < C_0$, then Eq. (5.2) will have eventually periodic cycles with period-2. The upcoming examples will remit these specific details and the patterns of the transient terms and the period-2 cycle when $x_0^2 < C_0$. The case when $x_0^2 > \frac{1}{C_1}$ will be studied later in this section.

Example 5.16 *Suppose that $C_0 C_1 < 1$ and*

$$[C_0 C_1] C_0 < x_0^2 < C_0. \tag{5.83}$$

Then Eq. (5.2) is eventually periodic with period-2 with two transient terms.

Solution: *Via (5.83) and by iterations we acquire the two transient terms in square brackets:*

$$\left[x_0, \frac{1}{x_0} \right], \tag{5.84}$$

and for all $n \geq 0$ the cognate period-2 pattern:

$$x_{2+2n} = \frac{C_0}{x_0} \quad \text{and} \quad x_{3+2n} = \frac{x_0}{C_0}. \tag{5.85}$$

Example 5.17 *Suppose that $C_0 C_1 < 1$ and*

$$[C_0 C_1]^2 C_0 < x_0^2 < [C_0 C_1] C_0. \tag{5.86}$$

Then Eq. (5.2) is eventually periodic with period-2 with five transient terms.

Solution: *Via (5.86) and by iterations we obtain the five transient terms in square brackets:*

$$\left[x_0, \frac{1}{x_0}, \frac{C_0}{x_0}, \frac{x_0}{C_0}, \frac{C_0}{x_0} \right], \tag{5.87}$$

and for all $n \geq 0$ and the cognate period-2 pattern:

$$x_{5+2n} = \frac{C_1 C_0}{x_0} \quad \text{and} \quad x_{6+2n} = \frac{x_0}{C_1 C_0}. \tag{5.88}$$

The upcoming sketch (Figure 5.46) renders (5.87) and (5.88) when $C_0 = 0.4$, $C_1 = 0.5$, $x_0 = 0.25$ and $x_1 = 4$:

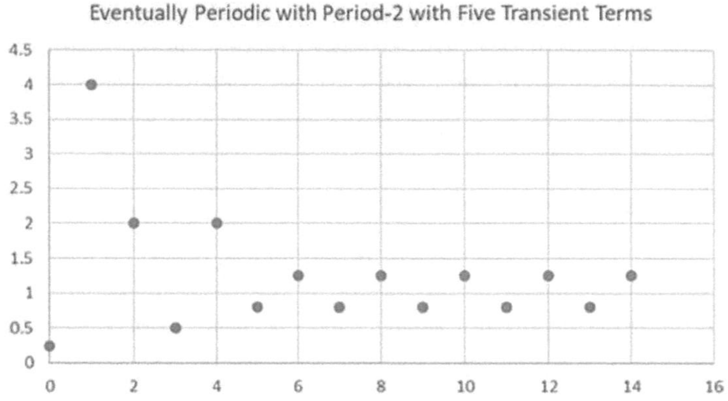

Eventually Periodic with Period-2 with Five Transient Terms

FIGURE 5.46: Eventually Periodic with Period-2 with Five Transient Terms.

The transient terms in Figure 5.46 emerge in **descending triangular-shape** *with the exception of the first two terms x_0 and x_1.*

Example 5.18 *Suppose that $C_0 C_1 < 1$ and*

$$[C_0 C_1]^3 C_0 < x_0^2 < [C_0 C_1]^2 C_0. \tag{5.89}$$

Then Eq. (5.2) is eventually periodic with period-2 with eight transient terms.

Solution: *Via (5.89) and by iterations we acquire the eight transient terms in square brackets:*

$$\left[x_0, \frac{1}{x_0}, \frac{C_0}{x_0}, \frac{x_0}{C_0}, \frac{C_0}{x_0}, \frac{C_0 C_1}{x_0}, \frac{x_0}{C_0 C_1}, \frac{C_0 C_1}{x_0} \right], \tag{5.90}$$

and for all $n \geq 0$ and the cognate period-2 pattern:

$$x_{8+2n} = \frac{C_0^2 C_1}{x_0} \quad \text{and} \quad x_{9+2n} = \frac{x_0}{C_0^2 C_1}. \tag{5.91}$$

The graph below (Figure 5.47) depicts (5.90) and (5.91) when $C_0 = 0.8$, $C_1 = 0.6$, $x_0 = 0.4$ and $x_1 = 2.5$:

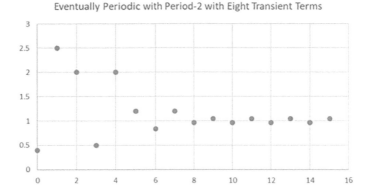

FIGURE 5.47: Eventually Periodic with Period-2 with Eight Transient Terms.

Analogous to Figure 5.46, the transient terms in Figure 5.47 are **descending triangular-shapes** *with the exception of the first two terms x_0 and x_1.*

Parallel to Figures 5.46 and 5.47, the upcoming sketch (Figure 5.48) evokes an eventually period-2 cycle when $C_0 = 0.8$, $C_1 = 0.9$, $x_0 = 0.5$ and $x_1 = 2$:

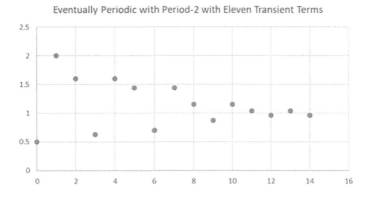

FIGURE 5.48: Eventually Periodic with Period-2 with Eleven Transient Terms.

Parallel with Figures 5.41, 5.6 and 5.42, Eq. (5.2) can also exhibit eventually constant solutions with $K = 1$ when $\max\{C_0, C_1\} \leq 1$. For instance, the cognate sketch (Figure 5.49) depicts an eventually constant solution with $K = 1$ with five transient terms when $C_0 = 0.5$, $C_1 = 0.4$, $x_0 = 0.2$ and $x_1 = 5$:

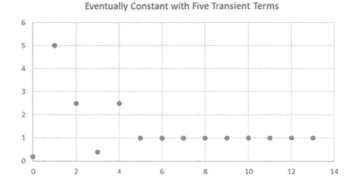

FIGURE 5.49: Eventually Constant with Five Transient Terms.

Now via (5.83), (5.86) and (5.89), there exists $k \geq 0$ such that:

$$[C_0C_1]^{k+1} < \frac{x_0^2}{C_0} < [C_0C_1]^k . \tag{5.92}$$

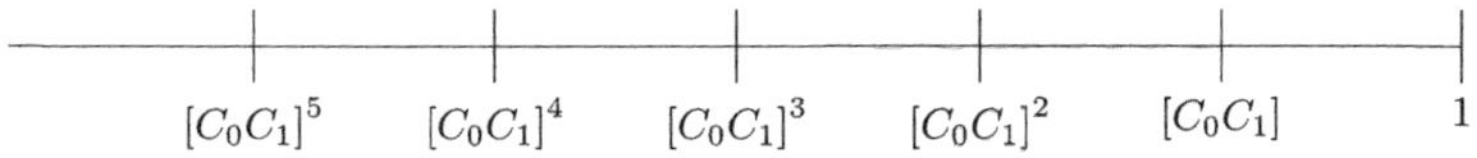

FIGURE 5.50: We choose $\frac{x_0^2}{C_0} \in \left[[C_0C_1]^{k+1}, [C_0C_1]^k\right]$ for some $k \geq 0$.

Note that (5.92) is analogous to (5.19) and Figure 5.50 is analogous to Figure 5.20. The upcoming theorem extends from Examples (5.16)–(5.18) and from Figures 5.46–5.48.

Theorem 5.13 *Suppose that $C_0C_1 < 1$ and (5.92) for some $k \geq 0$. Then Eq. (5.2) is eventually periodic with period-2 with $3k + 2$ transient terms with exactly one of the following period-2 patterns.*

(i) If $k \in [0, 2, \ldots]$, then for all $n \geq 0$:

$$x_{3k+2+2n} = \frac{C_0\,[C_0C_1]^{\frac{k}{2}}}{x_0} \quad \text{and} \quad x_{3k+3+2n} = \frac{x_0}{C_0\,[C_0C_1]^{\frac{k}{2}}}. \tag{5.93}$$

(ii) If $k \in [1, 3, \ldots]$, then for all $n \geq 0$:

$$x_{3k+2+2n} = \frac{[C_0C_1]^{\frac{k+1}{2}}}{x_0} \quad \text{and} \quad x_{3k+3+2n} = \frac{x_0}{[C_0C_1]^{\frac{k+1}{2}}}. \tag{5.94}$$

The proofs of (5.93) and (5.94) generalize from (5.85), (5.88) and (5.91). Analogous to (5.93) and (5.94), describing the transient terms' patterns will require two cases when $k \in [0, 2, \ldots]$ and $k \in [1, 3, \ldots]$. This will be left as an end-of-chapter exercise.

Now suppose that $x_0^2 > \frac{1}{C_1}$. We will show that Eq. (5.2) will be eventually periodic with period-2 with $3k$ transient terms ($k \in \mathbb{N}$). Then analogous to (5.22) and (5.25) in Examples (5.5) and (5.6) we will determine comparable intervals and analogous patterns.

Example 5.19 *Suppose that $C_0 C_1 < 1$ and*

$$\frac{1}{C_1} < x_0^2 < \frac{1}{C_1 [C_0 C_1]}. \tag{5.95}$$

Then Eq. (5.2) is eventually periodic with period-2 with three transient terms.

Solution: *Via (5.95) and by iterations we procure the three transient terms in square brackets:*

$$\left[x_0, \frac{1}{x_0}, x_0 \right], \tag{5.96}$$

and for all $n \geq 0$ the cognate period-2 pattern:

$$x_{3+2n} = C_1 x_0 \quad \text{and} \quad x_{4+2n} = \frac{1}{C_1 x_0}. \tag{5.97}$$

Example 5.20 *Suppose that $C_0 C_1 < 1$ and*

$$\frac{1}{C_1 [C_0 C_1]} < x_0^2 < \frac{1}{C_1 [C_0 C_1]^2}. \tag{5.98}$$

Then Eq. (5.2) is eventually periodic with period-2 with six transient terms.

Solution: *Via (5.98) and by iterations we obtain the six transient terms in square brackets:*

$$\left[x_0, \frac{1}{x_0}, x_0, C_1 x_0, \frac{1}{C_1 x_0}, C_1 x_0 \right], \tag{5.99}$$

and for all $n \geq 0$ the cognate period-2 pattern:

$$x_{6+2n} = C_0 C_1 x_0 \quad \text{and} \quad x_{7+2n} = \frac{1}{C_0 C_1 x_0}. \tag{5.100}$$

The consequent sketch (Figure 5.51) describes (5.99) and (5.100) with six **descending triangular-shaped** *transient terms when $C_0 = 0.8$, $C_1 = 0.5$, $x_0 = 3$ and $x_1 = \frac{1}{3}$:*

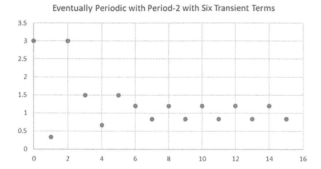

FIGURE 5.51: Eventually Periodic with Period-2 with Six Transient Terms.

Example 5.21 *Suppose that $C_0C_1 < 1$ and*

$$\frac{1}{C_1\left[C_0C_1\right]^2} < x_0^2 < \frac{1}{C_1\left[C_0C_1\right]^3}. \tag{5.101}$$

Then Eq. (5.2) is eventually periodic with period-2 with nine transient terms.

Solution: *Via (5.101) and by iterations we acquire the nine transient terms in square brackets:*

$$\left[\; x_0, \frac{1}{x_0}, x_0, C_1x_0, \frac{1}{C_1x_0}, C_1x_0, C_0C_1x_0, \frac{1}{C_0C_1x_0}, C_0C_1x_0 \;\right], \tag{5.102}$$

and for all $n \geq 0$ the cognate period-2 pattern:

$$x_{9+2n} = C_1\left[C_0C_1\right]x_0 \quad \text{and} \quad x_{10+2n} = \frac{1}{C_1\left[C_0C_1\right]x_0}. \tag{5.103}$$

The cognate graph (Figure 5.52) traces (5.102) and (5.103) with nine **descending triangular-shaped** *transient terms when $C_0 = 0.8$, $C_1 = 0.5$, $x_0 = 4$ and $x_1 = 0.25$:*

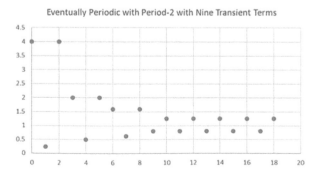

FIGURE 5.52: Eventually Periodic with Period-2 with Nine Transient Terms.

Eq. (5.2) can also exhibit eventually constant solutions with $K = 1$ when $\max\{C_0, C_1\} \leq 1$. The associated diagram (Figure 5.53) depicts an eventually constant solution with $K = 1$ with nine **descending triangular-shaped** transient terms when $C_0 = 0.8$, $C_1 = 0.5$, $x_0 = 5$ and $x_1 = 0.2$:

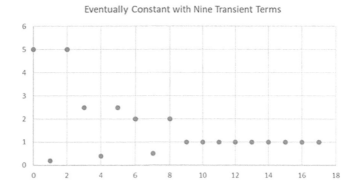

Eventually Constant with Nine Transient Terms

FIGURE 5.53: Eventually Constant with Nine Transient Terms.

Now via (5.95), (5.98) and (5.101), there exists $k \geq 0$ such that:

$$\frac{1}{[C_0 C_1]^k} < C_1 x_0^2 < \frac{1}{[C_0 C_1]^{k+1}}. \tag{5.104}$$

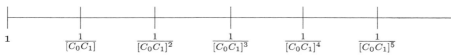

FIGURE 5.54: We choose $C_1 x_0^2 \in \left[\frac{1}{[C_0 C_1]^k}, \frac{1}{[C_0 C_1]^{k+1}} \right]$ for some $k \geq 0$.

Observe that (5.104) is similar to (5.42) and Figure 5.54 is analogous to Figure 5.17. The upcoming theorem extends from Examples (5.19)–(5.21) and from Figures 5.51 and 5.43.

Theorem 5.14 *Suppose that $C_0 C_1 < 1$ and (5.104) for some $k \geq 0$. Then Eq. (5.2) is eventually periodic with period-2 with $3k$ transient terms with exactly one of the following period-2 patterns.*

(i) If $k \in [0, 2, \ldots]$, then for all $n \geq 0$:

$$x_{3k+2n} = C_1 \left[C_0 C_1 \right]^{\frac{k}{2}} x_0 \quad \text{and} \quad x_{3k+1+2n} = \frac{1}{C_1 \left[C_0 C_1 \right]^{\frac{k}{2}} x_0}. \tag{5.105}$$

(ii) If $k \in [1, 3, \ldots]$, then for all $n \geq 0$:

$$x_{3k+2n} = \left[C_0 C_1 \right]^{\frac{k+1}{2}} x_0 \quad \text{and} \quad x_{3k+1+2n} = \frac{1}{\left[C_0 C_1 \right]^{\frac{k+1}{2}} x_0}. \tag{5.106}$$

The proofs of (5.105) and (5.106) generalize from (5.97), (5.100) and (5.103). Parallel to (5.105) and (5.106), expressing the transient terms' patterns will require two cases when $k \in [0, 2, \ldots]$ and $k \in [1, 3, \ldots]$. This will be left as an end-of-chapter exercise.

5.2.2   Eventually Periodic with Period-4

We will assume that $C_0 C_1 > 1$. Analogous to Example (5.15), the next example will portray the necessary and sufficient conditions for every solution of Eq. (5.2) to be periodic with period-4.

Example 5.22 *Suppose that $C_0 C_1 > 1$. Then Eq. (5.2) has period-4 cycles if and only if*

$$\frac{1}{C_1} < x_0^2 < C_0. \tag{5.107}$$

Therefore, via (5.107) if either $x_0^2 < \frac{1}{C_1}$ or $x_0^2 > C_0$, then Eq. (5.2) will have eventually periodic cycles with period-4. Analogous to Examples (5.16)–(5.18), the succeeding examples will remit these specific details and the patterns of the transient terms and the period-4 cycle when $x_0^2 > C_0$. The case when $x_0^2 < \frac{1}{C_1}$ will be examined later.

Example 5.23 *Suppose that $C_0 C_1 > 1$ and*

$$C_0 < x_0^2 < [C_0 C_1] C_0. \tag{5.108}$$

Then Eq. (5.2) is eventually periodic with period-4 with four transient terms.

Solution: *Via (5.108) and by iterations we acquire the four transient terms and for all $n \geq 0$ the cognate period-4 pattern:*

$$\begin{cases} x_{4+4n} = \frac{C_0}{x_0}, \\[2mm] x_{5+4n} = \frac{x_0}{C_0}, \\[2mm] x_{6+4n} = x_0, \\[2mm] x_{7+4n} = \frac{C_0 C_1}{x_0}. \end{cases} \tag{5.109}$$

The consequent diagram (Figure 5.55) depicts (5.109) with four transient terms when $C_0 = 1$, $C_1 = 2$, $x_0 = 1.2$ and $x_1 = \frac{5}{6}$:

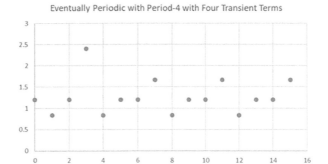

FIGURE 5.55: Eventually Periodic with Period-4 with Four Transient Terms.

Figure 5.55 portrays an **ascending step-shaped** *period-4 pattern.*

Example 5.24 *Suppose that $C_0C_1 > 1$ and*

$$C_0\left[C_0C_1\right] < x_0^2 < C_0\left[C_0C_1\right]^2. \tag{5.110}$$

Then Eq. (5.2) is eventually periodic with period-4 with seven transient terms.

Solution: *Via (5.110) and by iterations we acquire the seven transient terms and for all $n \geq 0$:*

$$\begin{cases} x_{7+4n} = \frac{C_0C_1}{x_0}, \\ x_{8+4n} = \frac{x_0}{C_0C_1}, \\ x_{9+4n} = \frac{x_0}{C_0}, \\ x_{10+4n} = \frac{C_0[C_0C_1]}{x_0}. \end{cases} \tag{5.111}$$

The next graph (Figure 5.56) describes (5.111) with seven transient terms when $C_0 = 2$, $C_1 = 3$, $x_0 = 5$ and $x_1 = 0.2$:

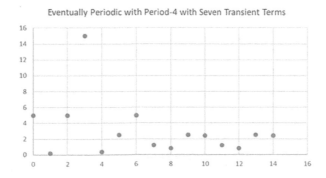

FIGURE 5.56: Eventually Periodic with Period-4 with Seven Transient Terms.

Example 5.25 *Suppose that $C_0 C_1 > 1$ and*

$$C_0 \left[C_0 C_1 \right]^2 < x_0^2 < C_0 \left[C_0 C_1 \right]^3. \tag{5.112}$$

Then Eq. (5.2) is eventually periodic with period-4 with ten transient terms.

Solution: *Via (5.112) and by iterations we obtain the eleven transient terms and for all $n \geq 0$:*

$$\begin{cases} x_{10+4n} = \dfrac{C_0 [C_0 C_1]}{x_0}, \\[2mm] x_{11+4n} = \dfrac{x_0}{C_0 [C_0 C_1]}, \\[2mm] x_{12+4n} = \dfrac{x_0}{C_0 C_1}, \\[2mm] x_{13+4n} = \dfrac{C_0 [C_0 C_1]^2}{x_0}. \end{cases} \tag{5.113}$$

Note that the first two terms of the period-4 cycles in (5.109), (5.111) and (5.113) are reciprocals of each other. In addition, (5.109) and (5.113) render an odd-parity pattern, while (5.111) describes an even-parity pattern that we will encounter in Theorem (5.15).

Now via (5.108), (5.110) and (5.112), there exists $k \in \mathbb{N}$ such that (Figure 5.57):

$$[C_0 C_1]^{k-1} < \frac{x_0^2}{C_0} < [C_0 C_1]^k. \tag{5.114}$$

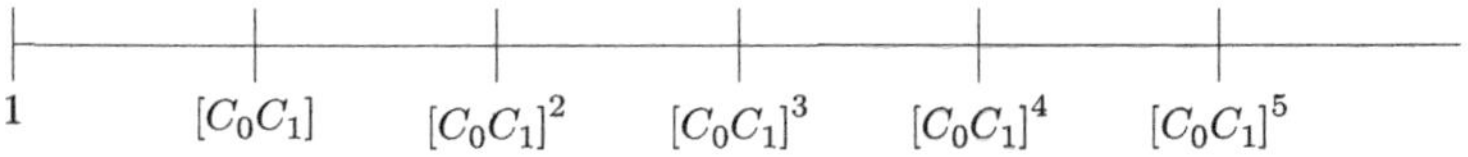

FIGURE 5.57: We choose $\frac{x_0^2}{C_0} \in \left[[C_0 C_1]^{k-1}, \ [C_0 C_1]^k \right]$ for some $k \in \mathbb{N}$.

The next theorem transitions from Examples (5.23), (5.24) and (5.25) and from (5.109), (5.111) and (5.113).

Theorem 5.15 *Suppose that $C_0 C_1 > 1$ and (5.114) for some $k \in \mathbb{N}$. Then Eq. (5.2) is eventually periodic with period-4 with $3k + 1$ transient terms with exactly one of the following period-4 patterns.*

(i) If $k \in [1, 3, \ldots]$, then for all $n \geq 0$:

$$\begin{cases} x_{3k+1+4n} = \dfrac{C_0 [C_0 C_1]^{\frac{k-1}{2}}}{x_0}, \\[3mm] x_{3k+2+4n} = \dfrac{x_0}{C_0 [C_0 C_1]^{\frac{k-1}{2}}}, \\[3mm] x_{3k+3+4n} = \dfrac{x_0}{[C_0 C_1]^{\frac{k-1}{2}}}, \\[3mm] x_{3k+4+4n} = \dfrac{[C_0 C_1]^{\frac{k+1}{2}}}{x_0}. \end{cases} \tag{5.115}$$

(ii) If $k \in [2, 4, \ldots]$, then for all $n \geq 0$:

$$
\begin{cases}
x_{3k+1+4n} = \dfrac{[C_0 C_1]^{\frac{k}{2}}}{x_0}, \\[2mm]
x_{3k+2+4n} = \dfrac{x_0}{[C_0 C_1]^{\frac{k}{2}}}, \\[2mm]
x_{3k+3+4n} = \dfrac{x_0}{C_0 [C_0 C_1]^{\frac{k-2}{2}}}, \\[2mm]
x_{3k+4+4n} = \dfrac{C_0 [C_0 C_1]^{\frac{k}{2}}}{x_0}.
\end{cases}
\tag{5.116}
$$

The proofs of (5.115) and (5.116) follow directly from (5.109), (5.111) and (5.113).

Next we will suppose that $x_0^2 < \frac{1}{C_1}$. In this case every solution of Eq. (5.2) is eventually periodic with period-4 with $3k$ transient terms ($k \in \mathbb{N}$). We will render some graphical examples of eventually periodic solutions of Eq. (5.2) with period-4 with $3k$ transient terms. The first graph (Figure 5.58) describes an eventually periodic cycle with period-4 with three **ascending transient terms** of Eq. (5.2) when $C_0 = 2$, $C_1 = 4$, $x_0 = 0.4$ and $x_1 = 2.5$:

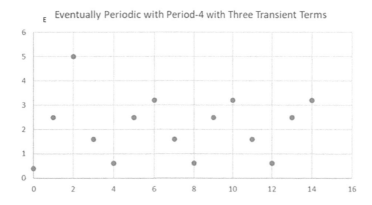

FIGURE 5.58: Eventually Periodic with Period-4 with Three Transient Terms.

The next sketch (Figure 5.59) portrays an eventually periodic cycle with period-4 with six transient terms of Eq. (5.2) when $C_0 = 2$, $C_1 = 4$, $x_0 = 0.1$ and $x_1 = 10$:

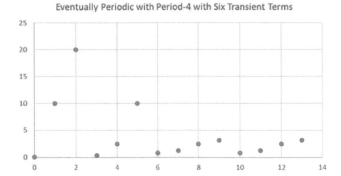

FIGURE 5.59: Eventually Periodic with Period-4 with Six Transient Terms.

Analogous to (5.114), there exists $k \in \mathbb{N}$ such that:

$$\frac{1}{[C_0 C_1]^k} < C_1 x_0^2 < \frac{1}{[C_0 C_1]^{k-1}}. \tag{5.117}$$

Via (5.117), obtaining patterns of periodic cycles and the transient terms will be left as an end-of-chapter exercise.

5.2.3   Eventually Periodic with Period-6

In this section we will assume that $C_0 C_1 = 1$. Parallel to the case when $C = 1$ of Eq. (5.3), our aim is to show that every solution of Eq. (5.2) is either periodic with period-6 or eventually periodic with period-6 with maximum three transient terms. We will render several assorted graphical examples.

The first two graphs portray period-6 cycles of Eq. (5.2) with various shapes depending on the values of C_0, C_1, x_0 and x_1. The sketch below (Figure 5.60) renders a period-6 cycle of Eq. (5.2) as **ascending and descending triangles** when $C_0 = 0.4$, $C_1 = 2.5$, $x_0 = 0.8$ and $x_1 = 1.25$:

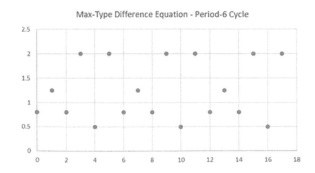

FIGURE 5.60: Ascending and Descending Triangular-Shaped Period-6 Cycle.

The upcoming diagram (Figure 5.61) portrays a period-6 cycle of Eq. (5.2) as **step-shaped and descending triangles** when $C_0 = 0.5$, $C_1 = 2$, $x_0 = 1$ and $x_1 = 1$:

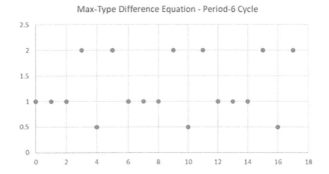

FIGURE 5.61: Step and Descending Triangular-Shaped Period-6 Cycle.

The upcoming graph (Figure 5.62) describes an eventually periodic cycle of Eq. (5.2) with period-6 with one transient term when $C_0 = 0.5$, $C_1 = 2$, $x_0 = 0.5$ and $x_1 = 1$:

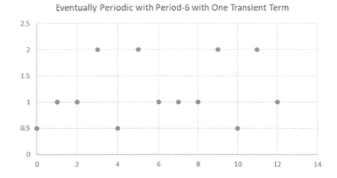

FIGURE 5.62: Eventually Periodic with Period-6 with One Transient Term.

Notice that Figure 5.62 depicts a similar pattern of the period-6 cycle in comparison with Figure 5.61. The upcoming diagram (Figure 5.63) traces a period-6 cycle of Eq. (5.2) with two descending transient terms when $C_0 = 2$, $C_1 = 0.5$, $x_0 = 2$ and $x_1 = 1.2$:

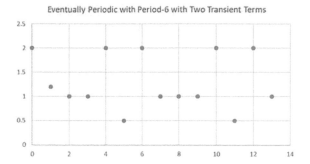

FIGURE 5.63: Eventually Periodic with Period-6 with Two Transient Terms.

The graph below (Figure 5.64) portrays a period-6 cycle of Eq. (5.2) with three ascending transient terms when $C_0 = 2.5$, $C_1 = 0.4$, $x_0 = 0.4$ and $x_1 = 1.5$:

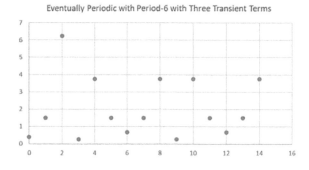

FIGURE 5.64: Eventually Periodic with Period-6 with Three Transient Terms.

Analogous to Theorems (5.7), (5.8), (5.9) and (5.10), our aim is to determine the necessary and sufficient conditions of Eq. (5.2) to be:

1. Periodic with period-6.

2. Eventually periodic with period-6 with one transient term.

3. Eventually periodic with period-6 with two transient terms.

4. Eventually periodic with period-6 with three transient terms.

This will be left as end-of-chapter exercises. We will retreat this chapter with additional studies on Eq. (5.2) with the cognate questions:

- When $\{C_n\}_{n=0}^{\infty}$ is a period-3 sequence:

  - Eq. (5.2) will have eventually periodic solutions with period-2 if $\max\{C_0, C_1, C_2\} < 1$;

- Eq. (5.2) will have eventually periodic solutions with period-3 if $\max\{C_0, C_1, C_2\} = 1$;
- Eq. (5.2) will have eventually periodic solutions with period-12 if $\min\{C_0, C_1, C_2\} > 1$;

- Periodic and Eventually Periodic cycles of Eq. (5.2) when $\{C_n\}_{n=0}^{\infty}$ is a period-k sequence, $(k \geq 4)$?

5.3 Chapter 5 Exercises

Consider the **Max-Type Δ.E.**:

$$x_{n+2} = \max\left\{\frac{1}{x_{n+1}}, \frac{C}{x_n}\right\}, \quad n = 0, 1, \ldots,$$

where $C > 0$ and $x_0, x_1 > 0$. In problems 1–10:

1. Let $C < 1$, $x_0 = C$ and $x_1 > 0$. Using computer observations, determine the number of transient terms of the eventually constant solution.

2. Let $C < 1$, $x_0 = C^2$ and $x_1 > 0$. Using computer observations, determine the number of transient terms of the eventually constant solution.

3. Let $C < 1$, $x_0 = C^3$ and $x_1 > 0$. Using computer observations, determine the number of transient terms of the eventually constant solution.

4. Let $C < 1$, $x_0 = C^4$ and $x_1 > 0$. Using computer observations, determine the number of transient terms of the eventually constant solution.

5. Using exercises 1–4, let $C < 1$, $x_0 = C^k$ and $x_1 > 0$ ($k \in \mathbb{N}$). Using computer observations, determine the number of transient terms of the eventually constant solution.

6. Let $C < 1$, $x_0 = \frac{1}{C}$ and $x_1 > 0$. Using computer observations, determine the number of transient terms of the eventually constant solution.

7. Let $C < 1$, $x_0 = \frac{1}{C^2}$ and $x_1 > 0$. Using computer observations, determine the number of transient terms of the eventually constant solution.

8. Let $C < 1$, $x_0 = \frac{1}{C^3}$ and $x_1 > 0$. Using computer observations, determine the number of transient terms of the eventually constant solution.

9. Let $C < 1$, $x_0 = \frac{1}{C^4}$ and $x_1 > 0$. Using computer observations, determine the number of transient terms of the eventually constant solution.

10. Using exercises 6–9, let $C < 1$, $x_0 = \frac{1}{C^k}$ and $x_1 > 0$ ($k \in \mathbb{N}$). Using computer observations, determine the number of transient terms of the eventually constant solution.

Consider the **Max-Type** Δ.**E.**:

$$x_{n+2} = \max\left\{\frac{1}{x_{n+1}}, \frac{C_n}{x_n}\right\}, \quad n = 0, 1, \ldots .$$

In problems 11–18, suppose that $C_0, C_1 \in (0, 1)$:

11. Let $C_0 = 0.5$, $C_1 = 0.4$, $x_0 = 0.5$ and $x_1 = 2$. Using computer observations, determine the number of transient terms of the eventually constant solution.

12. Let $C_0 = 0.5$, $C_1 = 0.4$, $x_0 = 0.25$ and $x_1 = 4$. Using computer observations, determine the number of transient terms of the eventually constant solution.

13. Let $C_0 = 0.5$, $C_1 = 0.4$, $x_0 = 0.2$ and $x_1 = 5$. Using computer observations, determine the number of transient terms of the eventually constant solution.

14. Let $C_0 = 0.8$, $C_1 = 0.5$, $x_0 = 2.5$ and $x_1 = 0.4$. Using computer observations, determine the number of transient terms of the eventually constant solution.

15. Let $C_0 = 0.8$, $C_1 = 0.5$, $x_0 = 5$ and $x_1 = 0.2$. Using computer observations, determine the number of transient terms of the eventually constant solution.

16. Let $C_0 = 0.5$, $C_1 = 0.4$, $x_0 = 0.1$ and $x_1 = 10$. Using computer observations, determine the number of transient terms of the eventually constant solution.

17. Let $C_0 = 0.8$, $C_1 = 0.5$, $x_0 = 6.25$ and $x_1 = 0.16$. Using computer observations, determine the number of transient terms of the eventually constant solution.

18. Let $C_0 = 0.5$, $C_1 = 1$, $x_0 = 0.0625$ and $x_1 = 4$. Using computer observations, determine the number of transient terms of the eventually constant solution.

In problems 19–22, suppose that $C_0 C_1 > 1$:

19. Let $\frac{1}{[C_0 C_1]^2} < C_1 x_0^2 < \frac{1}{[C_0 C_1]}$. Determine the pattern of the transient terms and the period-4 pattern.

20. Let $\frac{1}{[C_0 C_1]^3} < C_1 x_0^2 < \frac{1}{[C_0 C_1]^2}$. Determine the pattern of the transient terms and the period-4 pattern.

21. Let $\frac{1}{[C_0 C_1]^4} < C_1 x_0^2 < \frac{1}{[C_0 C_1]^3}$. Determine the pattern of the transient terms and the period-4 pattern.

22. Let $\frac{1}{[C_0 C_1]^5} < C_1 x_0^2 < \frac{1}{[C_0 C_1]^4}$. Determine the pattern of the transient terms and the period-4 pattern.

Chapter 6

Appendices

6.1 Patterns of Sequences

1. Linear Patterns:

$$1,\ 2,\ 3,\ 4,\ 5,\ 6,\ 7,\ \ldots\ =\ \{n\}_{n=1}^{\infty}$$
$$2,\ 4,\ 6,\ 8,\ 10,\ 12,\ 14,\ \ldots\ =\ \{2n\}_{n=1}^{\infty}$$
$$1,\ 3,\ 5,\ 7,\ 9,\ 11,\ 13,\ \ldots\ =\ \{2n+1\}_{n=0}^{\infty}$$
$$3,\ 6,\ 9,\ 12,\ 15,\ 18,\ 21,\ \ldots\ =\ \{3n\}_{n=1}^{\infty}$$

2. Quadratic Patterns:

$$1,\ 4,\ 9,\ 16,\ 25,\ 36,\ 49,\ldots\ =\ \{n^2\}_{n=1}^{\infty}$$
$$4,\ 16,\ 36,\ 64,\ 100,\ 144,\ 196,\ldots\ =\ \{(2n)^2\}_{n=1}^{\infty}$$
$$1,\ 9,\ 25,\ 49,\ 81,\ 121,\ 169,\ldots\ =\ \{(2n-1)^2\}_{n=1}^{\infty}$$

3. Geometric Patterns:

$$1,\ r,\ r^2,\ r^3,\ r^4,\ r^5,\ r^6,\ldots\ =\ \{r^n\}_{n=0}^{\infty}$$
$$2,\ 4,\ 8,\ 16,\ 32,\ 64,\ 128,\ldots\ =\ \{2^n\}_{n=1}^{\infty}$$
$$3,\ 9,\ 27,\ 81,\ 243,\ 729,\ldots\ =\ \{3^n\}_{n=1}^{\infty}$$

6.2 Alternating Patterns of Sequences

1. Linear Alternating Patterns:

$$1,\ -2,\ 3,\ -4,\ 5,\ -6,\ 7,\ \ldots\ =\ \{(-1)^{n+1}\,n\}_{n=1}^{\infty}$$
$$-1,\ 2,\ -3,\ 4,\ -5,\ 6,\ -7,\ \ldots\ =\ \{(-1)^{n}\,n\}_{n=1}^{\infty}$$
$$1,\ -3,\ 5,\ -7,\ 9,\ -11,\ 13,\ \ldots\ =\ \{(-1)^{n}\,[2n+1]\}_{n=0}^{\infty}$$
$$-1,\ 3,\ -5,\ 7,\ -9,\ 11,\ -13,\ \ldots\ =\ \{(-1)^{n+1}\,[2n+1]\}_{n=0}^{\infty}$$

2. **Quadratic Alternating Patterns:**

$$1,\ -4,\ 9,\ -16,\ 25,\ -36,\ 49,\ldots\ =\ \{(-1)^{n+1}\ n^2\}_{n=1}^{\infty}$$
$$-1,\ 4,\ -9,\ 16,\ -25,\ 36,\ -49,\ldots\ =\ \{(-1)^{n}\ n^2\}_{n=1}^{\infty}$$

3. **Alternating Geometric Patterns:**

$$1,\ -r,\ r^2,\ -r^3,\ r^4,\ -r^5,\ r^6,\ldots\ =\ \{(-1)^{n}\ r^{n}\}_{n=0}^{\infty}$$
$$-1,\ r,\ -r^2,\ r^3,\ -r^4,\ r^5,\ -r^6,\ldots\ =\ \{(-1)^{n+1}\ r^{n}\}_{n=0}^{\infty}$$

6.3   Finite Series

$$1 + 2 + 3 + 4 + 5 + 6 + \ldots + n = \sum_{i=1}^{n} i = \frac{n[n+1]}{2}.$$

$$1 + 3 + 5 + 7 + 9 + 11 + \ldots + [2n-1] = \sum_{i=1}^{n} (2i-1) = n^2.$$

$$1 + 4 + 9 + 16 + 25 + 36 + \ldots + n^2 = \sum_{i=1}^{n} i^2 = \frac{n[n+1][2n+1]}{6}.$$

$$1 \cdot 2 + 2 \cdot 3 + 3 \cdot 4 + 4 \cdot 5 + \ldots + n \cdot [n+1] = \sum_{i=1}^{n} i \cdot [i+1] = \frac{n[n+1][n+2]}{3}.$$

$$\frac{1}{1 \cdot 2} + \frac{1}{2 \cdot 3} + \frac{1}{3 \cdot 4} + \frac{1}{4 \cdot 5} + \ldots + \frac{1}{n \cdot [n+1]} = \sum_{i=1}^{n} \frac{1}{i \cdot [i+1]} = \frac{n}{n+1}.$$

$$1 + r + r^2 + r^3 + r^4 + r^5 + \ldots + r^n = \sum_{i=0}^{n} r^i = \frac{1-r^{n+1}}{1-r}.$$

$$1 \cdot 2^0 + 2 \cdot 2^1 + 3 \cdot 2^2 + \ldots + n \cdot 2^{n-1} = \sum_{i=1}^{n} i \cdot 2^{i-1} = [n-1]2^n + 1.$$

$$\binom{n}{0} + \binom{n}{1} + \binom{n}{2} + \ldots + \binom{n}{n-1} + \binom{n}{n} = \sum_{i=0}^{n} \binom{n}{i} = 2^n.$$

6.4   Convergent Infinite Series

$$1 + r + r^2 + r^3 + r^4 + r^5 + \ldots + = \sum_{n=0}^{\infty} r^n = \frac{1}{1-r},\ |r| < 1.$$

$$\frac{1}{1 \cdot 2} + \frac{1}{2 \cdot 3} + \frac{1}{3 \cdot 4} + \frac{1}{4 \cdot 5} + \ldots + = \sum_{n=1}^{\infty} \frac{1}{n[n+1]} = 1.$$

$$2 + \frac{1}{2} + \frac{1}{6} + \frac{1}{24} + \frac{1}{120} + \frac{1}{720} + \ldots + = \sum_{n=0}^{\infty} \frac{1}{n!} = e.$$

$$1 - \frac{1}{2} + \frac{1}{3} - \frac{1}{4} + \frac{1}{5} - \frac{1}{6} + \ldots + = \sum_{n=1}^{\infty} \frac{(-1)^{n+1}}{n} = Ln[2].$$

$$1 + \frac{1}{4} + \frac{1}{9} + \frac{1}{16} + \frac{1}{25} + \frac{1}{36} + \ldots + = \sum_{n=1}^{\infty} \frac{1}{n^2} = \frac{\pi^2}{6}.$$

6.5   Periodicity and Modulo Arithmetic

Period-2 sequence $\{A_n\}_{n=0}^{\infty}$ and the following period-2 pattern:

$$A_0,\ A_1,\ A_0,\ A_1,\ \dots.$$

Period-2 sequence $\{A_n\}_{n=0}^{\infty}$ and the following period-2 pattern:

$$\frac{A_0A_1 - 1}{1 + A_0},\ \frac{A_0A_1 - 1}{1 + A_1},\ \frac{A_0A_1 - 1}{1 + A_0},\ \frac{A_0A_1 - 1}{1 + A_1},\ \dots.$$

Period-3 sequence $\{A_n\}_{n=0}^{\infty}$ and the following period-3 pattern:

$$A_0,\ A_1,\ A_2,\ A_0,\ A_1,\ A_2,\ \dots.$$

Period-3 sequence $\{A_n\}_{n=0}^{\infty}$ and the following period-3 pattern:

$$\frac{A_0A_1}{A_0A_1A_2 + 1},\ \frac{A_1A_2}{A_0A_1A_2 + 1},\ \frac{A_2A_0}{A_0A_1A_2 + 1},\ \dots.$$

Period-4 sequence $\{A_n\}_{n=0}^{\infty}$ and the following period-4 pattern:

$$A_0,\ A_1,\ A_2,\ A_3,\ A_0,\ A_1,\ A_2,\ A_3,\ \dots.$$

Period-4 sequence $\{A_n\}_{n=0}^{\infty}$ and the following period-4 pattern:

$$\frac{A_0 + A_1 + A_2}{2},\ \frac{A_1 + A_2 + A_3}{2},\ \frac{A_2 + A_3 + A_0}{2},\ \frac{A_3 + A_0 + A_1}{2},\ \dots.$$

6.5.1   Alternating Periodicity

Period-2 sequence $\{A_n\}_{n=0}^{\infty}$ and the following period-4 pattern:

$$A_0,\ A_1,\ -A_0,\ -A_1,\ \dots.$$

Period-3 sequence $\{A_n\}_{n=0}^{\infty}$ and the following period-6 pattern:

$$A_0,\ A_1,\ A_2,\ -A_0,\ -A_1,\ -A_2,\dots.$$

Period-2 sequence $\{A_n\}_{n=0}^{\infty}$ and the following period-2 pattern:

$$\frac{A_0A_1}{A_0 + A_1 + 1},\ \frac{-A_0A_1}{A_0 + A_1 + 1},\ \dots.$$

Period-2 sequence $\{A_n\}_{n=0}^{\infty}$ and the following period-2 pattern:

$$\frac{A_0 - A_1}{A_0A_1 + 1},\ \frac{A_1 - A_0}{A_0A_1 + 1},\ \dots.$$

Period-2 sequence $\{A_n\}_{n=0}^{\infty}$ and the following period-4 pattern:

$$\frac{A_0}{A_0 A_1 + 1}, \ \frac{A_1}{A_0 A_1 + 1}, \ \frac{-A_0}{A_0 A_1 + 1}, \ \frac{-A_1}{A_0 A_1 + 1}, \ \ldots.$$

Period-2 sequence $\{A_n\}_{n=0}^{\infty}$ and the following period-4 pattern:

$$\frac{A_0 + A_1}{A_0 A_1 + 1}, \ \frac{A_0 - A_1}{A_0 A_1 + 1}, \ \frac{-[A_0 + A_1]}{A_0 A_1 + 1}, \ \frac{A_1 - A_0}{A_0 A_1 + 1}, \ \ldots.$$

6.6   Patterns as an Initial Value Problem

1. $1, 2, 3, 4, 5, \ldots.$

$$\begin{cases} x_{n+1} = x_n + 1, & n = 0, 1, \ldots, \\ x_0 = 1. \end{cases}$$

2. $2, 4, 6, 8, 10, \ldots.$

$$\begin{cases} x_{n+1} = x_n + 2, & n = 0, 1, \ldots, \\ x_0 = 2. \end{cases}$$

3. $k, 2k, 3k, 4k, 5k, \ldots.$

$$\begin{cases} x_{n+1} = x_n + k, & n = 0, 1, \ldots, \\ x_0 = k. \end{cases}$$

4. $\sum_{i=1}^{n} i.$

$$\begin{cases} x_{n+1} = x_n + (n + 2), & n = 0, 1, \ldots, \\ x_0 = 1. \end{cases}$$

5. $1, 2, 4, 8, 16, \ldots.$

$$\begin{cases} x_{n+1} = 2x_n, & n = 0, 1, \ldots, \\ x_0 = 1. \end{cases}$$

6. $1, 3, 9, 27, 81, \ldots.$

$$\begin{cases} x_{n+1} = 3x_n, & n = 0, 1, \ldots, \\ x_0 = 1. \end{cases}$$

7. $1, k, k^2, k^3, k^4, \ldots.$

$$\begin{cases} x_{n+1} = kx_n, & n = 0, 1, \ldots, \\ x_0 = 1. \end{cases}$$

8.

$$n! = \begin{cases} 1 & \text{if } n = 0, \\ \prod_{i=1}^{n} i & \text{if } n \in \mathbb{N}. \end{cases}$$

$$\begin{cases} x_{n+1} = (n+1)x_n, & n = 0, 1, \ldots, \\ x_0 = 1. \end{cases}$$

9. 1, 4, 9, 16, 25, ...

$$\begin{cases} x_{n+1} = x_n + (2n+3), & n = 0, 1, \ldots, \\ x_0 = 1. \end{cases}$$

10. 1, 9, 25, 49, 81, ...

$$\begin{cases} x_{n+1} = x_n + 8(n+1), & n = 0, 1, \ldots, \\ x_0 = 1. \end{cases}$$

11. $\sum_{i=1}^{n} 2^i$.

$$\begin{cases} x_{n+1} = x_n + 2^{n+1}, & n = 0, 1, \ldots, \\ x_0 = 1. \end{cases}$$

12. $\sum_{i=1}^{n} 3^i$.

$$\begin{cases} x_{n+1} = x_n + 2 \cdot 3^n, & n = 0, 1, \ldots, \\ x_0 = 1. \end{cases}$$

6.7 Specific Periodic Patterns

1. Period-2 pattern of Linear Δ.E. $x_{n+1} = -x_n$:

$$x_0, \ -x_0, \ \ldots.$$

2. Period-2 pattern of Linear Δ.E. $x_{n+1} = -x_n + b$:

$$x_0, \ -x_0 + b, \ \ldots.$$

3. Period-k pattern of Linear Δ.E. $x_{n+1} = a_n x_n$, where $\{a_n\}_{n=0}^{\infty}$ is a period-k sequence ($k \geq 2$) and $P = \prod_{i=0}^{k-1} a_i$:

$$x_{kn+k-j} = \left[\prod_{i=0}^{j-1} a_i \right] P^n x_0 \quad \text{for all} \quad j \in [0, 1, \ldots, k-1].$$

4. Period-k pattern of Linear Δ.E. $x_{n+1} = x_n + b_n$, where $\{b_n\}_{n=0}^{\infty}$ is a period-k sequence ($k \geq 2$) and $S = \sum_{i=0}^{k-1} b_i$:

$$x_{kn+k-j} = x_0 + \left[\sum_{i=0}^{j-i} b_i \right] + Sn \quad \text{for all} \quad j \in [0, 1, \ldots, k-1].$$

5. Period-(2k+1) pattern of Linear Δ.E. $x_{n+1} = -x_n + b_n$, where $\{b_n\}_{n=0}^{\infty}$ is a period-(2k+1) sequence ($k \in \mathbb{N}$):

$$x_j = \frac{\sum_{i=1}^{2k+1} (-1)^{i+1} b_{i+j-1}}{2} \quad \text{for all} \quad j \in [0, 1, \ldots, 2k].$$

6. Period-2m ($m \in \mathbb{N}$) pattern of Riccati Δ.E. $x_{n+m} = \frac{A}{x_n}$:

$$x_0, \ x_1, \ \ldots, \ x_{m-1}, \ \frac{A}{x_0}, \ \frac{A}{x_0}, \ \ldots, \ \frac{A}{x_{m-1}}.$$

7. Period-(2k+1) ($m \in \mathbb{N}$) pattern of Riccati Δ.E. $x_{n+1} = \frac{A_n}{x_n}$, where $\{A_n\}_{n=0}^{\infty}$ is a period-(2k+1) sequence ($k \in \mathbb{N}$):

$$x_j = \sqrt{\frac{\prod_{i=1}^{k+1} A_{2i-2+j}}{\prod_{i=1}^{k} A_{2i-1+j}}} \quad \text{for all} \quad j \in [0, 1, \ldots, 2k].$$

8. Period-p ($p \geq 2$) pattern of Tent-Map:

$$x_i = \frac{2^{i+1}}{2^p + 1} \quad \text{for all} \quad i \in [0, 1, 2, \ldots, p-1].$$

Bibliography

[1] Amleh, A.M., Hoag, J., Ladas, G., A difference equation with eventually periodic solutions. *Computers & Mathematics with Applications* **36**, 401–404 (1998).

[2] Anisimova, A., Avotina, M., Bula, I., Periodic orbits of single neuron models with internal decay rate $0 < \beta \leq 1$. *Mathematical Modelling and Analysis* **18**, 325–345 (2013).

[3] Anisimova, A., Avotina, M., Bula, I., Periodic and chaotic orbits of a neuron model. *Mathematical Modelling and Analysis* **20**, 30–52 (2015).

[4] Briden, W.J., Grove, E.A., Ladas, G., McGrath, L.C., On the non-autonomous equation $x_{n+1} = max\{\frac{A_n}{x_n}, \frac{B_n}{x_{n-1}}\}$. *Proceedings of the Third International Conference on Difference Equations and Applications.* September 1–5, 1997, Taipei, Taiwan, Gordon and Breach Science Publishers, 49–73 (1999).

[5] Briden, W.J., Ladas, G., Nesemann, T., On the recursive sequence $x_{n+1} = max\left\{\frac{1}{x_n}, \frac{A_n}{x_{n-1}}\right\}$, *Journal of Difference Equations and Applications* **5**, 491–494 (1999).

[6] Bula, I., Radin, M.A., Wilkins, N., Neuron model with a period three internal decay rate. *Electronic Journal of Qualitative Theory of Differential Equations* (EJQTDE) **46**, 1–19 (2017).

[7] Bula, I., Radin, M.A., Periodic orbits of a neuron model with periodic internal decay rate. *Applied Mathematics and Computation* **266**, 293–303 (2015).

[8] Kent, C.M., Kustesky, M. , Nguyen, A.Q. , Nguyen, B.V., Eventually periodic solutions of $x_{n+1} = max\left\{\frac{A_n}{x_n}, \frac{B_n}{x_{n-1}}\right\}$, With Period Two Cycle Parameters. *Dynamics of Continous, Discrete and Impulsive Systems, Series A* **10**, 33–40 (2003).

[9] Kocic, V.L., Darensburg, T., On the Discrete Model of West Nile-Like Epidemic. *Proceedings of Dynamic Systems and Applications*, 4(2004), 358–366.

[10] Ladas, G., On the recursive sequence $x_{n+1} = max\left\{\frac{A_0}{x_n}, \ldots, \frac{A_k}{x_{n-k}}\right\}$. *Journal of Difference Equations and Applications* **2**(2), 339–341 (1996).

[11] Orlova, O., Radin, M., *University level teaching styles with high school students and international teaching and learning.* International Scientific Conference "Society, Integration, Education" (2018).

[12] Radin, M., Riashchenko, V., Effective Pedagogical Management as a road to successful international teaching and learning. *Forum Scientiae Oeconomia* **5**(4), 71–84 (2017).

[13] Williamson, M., The analysis of discrete time cycles. In Usher, M.B., Williamson, M.H., eds. *Ecological stability*, 1974, Chapman and Hall, 17–33.

[14] Zhou, Z., Wu, J., Stable periodic orbits in nonlinear discrete-time neural networks with delayed feedback. *Computers and Mathematics with Applications* **45**, 935–942 (2003).

[15] Zhou, Z., Periodic orbits on discrete dynamical systems. *Computers and Mathematics with Applications* **45**, 1155–1161 (2003).

[16] Zhu, H., Huang, L., Dynamics of a class of nonlinear discrete-time neural Networks. *Computers and Mathematics with Applications* **48**, 85–94 (2004).

[17] Yuan, Z., Huang, L., Chen, Y., Convergence and periodicity of solutions for a Discrete-Time Network Model of Two Neurons. *Mathematical and Computer Modelling* **35**, 941–950 (2002).

[18] Yuan, Z., Huang, L., All solutions of a class of discrete-time systems are eventually periodic. *Applied Mathematics and Computation* **158**, 537–546 (2004).

Index